Wild Among Us

True adventures of a female wildlife photographer
who stalks bears, wolves, mountain lions, wild horses and other elusive wildlife

PAT TOTH-SMITH

DEDICATION

To Andy Smith, my husband, whose support and editing assistance made this book possible, and to my daughter, Alia, for her sacrifices in doing without a mom during the writing of this book and my time spent out in the field; to my departed father Bill Toth, for his wildlife games and obsession with nature; and lastly, to my departed mother, Helen Toth, who endlessly photographed me and encouraged me in the arts. I carry her and God with me on all my adventures.

ACKNOWLEDGEMENTS

I would like to thank Sally Smith for her editing skills and support, and Jerry Smith and Larry Greene for their helpful support and knowledge. Thanks also to my many family members and friends who supported my efforts: Joan Brozovich, Billy Toth, Charlie Weltner, Jack Barsnica, Thea Dahlberg, my beloved Choici, Debbie Obst, Fred Obst, Dana Obst, Alysha Obst, Angela Sledzik, Lori Sledzik-Bein, Joe Bein, Jade and Abby Warner, Lynda Browne, Micheal Browne, Lia and William Sagarese; Micheal Jr., Aidan, and Eva Browne; Tommy Toth, Joanne Toth, Helen Breuer, Martin Smith, Ellery Smith, Cole and Quinn Smith, Mary Jane Niven, Gerry Dollard, Dave Niven, John Niven, Kristine Hanna, Greg Lomax, Davis and Miranda Lomax, Joyce Hanna, Katie Dickson, Tom Dickson, Ben and Sam Marks, and Allison Dickson. Thanks also to my dear friends: Donna Mleczek, Doris Chapman, Ofelia Delgadillo, Andrea Wilson, Cindy, Kurt, Marisol Pena, Sandy Rarick, my best friend Sophie Robicheau, Phillip Robicheau, Ashley and Phillip Jr. Robicheau, Paul Greenley, Murphy Anderson, Helen Anderson, Jeff Kitzes, Sheridan Adams, Therese Babineau, Pat Kirch, Mary Silvia, John Kunze, Hugh Robinson, Kevin Wright at Avant; Leah, Kenny and Kallie Cook; Linda, Gilbert, and Amalia Von Stuben; and Neil Salkind.

ISBN: 978-0-9892513-3-4

www.tothsmithphotography.com

Graphic production: Nina Zurier

CONTENTS

As I walked slowly along the trail with my friend, my heart pounded in my chest, my head was bowed submissively, and my body sweated profusely. I wanted desperately to run, but I knew I couldn't and shouldn't. We were being stalked by a five-hundred-pound-plus grizzly bear and her cub.

I had cajoled another female artist, Sandy into joining me on this weeklong photo trip to Katmai National Park, Alaska. When we started out, we were excited. It would be a great photo adventure that we could use for our art and gallery exhibits. But things had suddenly gone very wrong.

We had planned to photograph the annual salmon run where thirty to fifty grizzly bears would gather to eat salmon at a waterfall on the Brooks River. We were staying at Brook's Camp, which consisted of a lodge and restaurant, overnight rooms, a visitor center, a campground area, and a ranger station located at the mouth of the Brooks River. It was an idyllic setting surrounded by a semi-dense forest with scattered grassy marsh areas.

After receiving our bear orientation, we were ready to begin a twenty-minute hike to the campground, where we planned to set up our tent.

The campground is about a quarter of a mile from the Brook's enclave of buildings. The trail to it runs along Kaknek Lake; it's a dirt path forty feet into the woods that parallels the beach. I had learned on a previous visit that one safeguard against coming across a bear on the trail was to first walk down to the lake and scan the beach for any bears that might make their way into the woods and across the trail. This time, when I walked to the beachfront, I surprised a grizzly bear and her cub. When she saw me, she growled. I backed away slowly, my head hunched over to convey to the mother bear that we weren't a threat to her or her cub. After what seemed like an eternity, I finally made my way back to the trailhead and Sandy. But the grizzly mom still stood thirty feet away, staring at us. My vision clouded, and it all seemed surreal.

Sandy and I began making our way to the campground, while the grizzly walked parallel to us along the beach. As we proceeded slowly, step by step, I could see the unhappy bear in the corner of my eye staring at us the whole time. This "stalk and walk" process seemed to go on forever. The palpable tension and our hyper-vigilance lasted for more than twenty minutes, and the comfort I usually found in being in the woods took on an ominous, eerie feel. I expected the bear to cross the short distance to us at any moment.

When we at last reached the campground, we scurried into the electrified fenced-in-area and breathed a sigh of relief. Like steers being corralled into their pen, we settled in for the night. Our bear-tormentor stayed outside the fence for an hour.

As I began to relax, I reflected about my fears: *"My place in the Alaskan food chain is crystal clear. This experience of being stalked has shaken me to my core. I'll be facing this same bear that just growled at me menacingly for the rest of the week. I feel humbled, scared, and vulnerable, and a huge part of me wants to ensconce myself in my little tent within this protected containment area for the rest of my time here."*

I prayed for insight about how to be with these bears for the next five days without incurring their wrath. In the security and comfort of the fenced-in fortress and my tent, I reflected on the harrowing walk in the woods that Sandy and I had just experienced. I thought about my life. I kept coming back to a central question: how had a New Jersey girl like me, who had hung out at the beach and in malls as a young adult, ever become a professional wildlife and nature photographer with an overwhelming curiosity about bears and other dangerous wildlife?

My early childhood experiences had sparked an interest in wildlife and photography. Growing up in New Jersey, I had limited exposure to wildlife. My father loved to hunt, so I was more used to seeing dead wild animals than live ones. It was not unusual for a deer corpse to be hanging from our garage ceiling at hunting season. An avid sportsman who couldn't wait for the different hunting seasons, my father had hunting dogs, camouflage outfits, all types of hiking and camping gear, and a cabinet full of rifles, cross bows and other weapons. I got used to eating venison. My father seemed giddy around wildlife and showed a playful side I rarely saw. We made a game of searching out wildlife, especially deer, as we drove around rural New Jersey. His enthusiasm spread to my brothers, sisters, and me and we competed with one another to find the next deer. As a middle child, I often felt marginalized but sometimes when I spotted a deer I rose in everyone's estimation, or so I thought. Thus began my lifelong interest in stalking animals.

Some other childhood experiences of the wild came from family trips to Silver Lake, a semi-wooded lake in upstate New Jersey. We initially used large tents for camping and later on an RV. We found Silver Lake by accident, when a ranger at a nearby state park recommended it for swimming. It was privately owned and had a swimming area with a raft to swim out to and boats to rent for fishing. The woman who owned the land and all the concessions on it was named Marjorie. She was tall and had a gray helmet-like pageboy haircut. She was gruff in temperament, had little use for words, and drove around in a pink Cadillac. Marjorie also owned property on the other side of the lake where she grazed cattle; this grazing land doubled as a campground. There were no designated campsites, no hookups, no picnic tables or fire pits, only a couple of outhouses and a water spigot. The entrance had no signs, only a rusted, locked gate, which was hard to find. One got the impression that a visitor could either take it or leave it, and it made no difference to Marjorie. But Dad was excited about camping next to the lake and jumped at the opportunity to rent a site.

Camping in tents in the middle of a cow pasture was an experience. The cattle would sniff around our tents and leave a few foul surprises in the night. Not surprisingly, there were few other campers besides my large extended family; my brothers, sisters, cousins, and I had the run of the place. We climbed trees and watched the cattle as they grazed below us, swam in the lake, and hiked out to a point on the property. In the evening, we occasionally snuck into a private summer camp on the lake and, jumped off their dock into the water. In the daytime we spied on them. It seemed like our own private lakefront property and we kids knew every inch of it.

As I got older I abandoned Silver Lake for the shopping mall and the Jersey shore's beaches and boardwalks. Sunbathing, swimming, and shopping became my new interests, with running thrown in to keep myself fit and in shape for all the shopping.

My early photographic experiences involved my mom. She was an avid amateur photographer. With her film camera and a primitive video camera she stalked her favorite subject—me. She would pose me in front of everything. My job was to crack a smile and brace for the vision-blinding flash. My parents' photo albums brimmed with my washed-out, smiling face. My mom found everything my four siblings and I did interesting, and like a journalistic documentarian, she set about recording our every move. Our household became a movie set, with my mom as make-up artist, camera crew, and director. These early experiences were the foundation for what would become my life's work, photography.

It wasn't until I moved to California in my late twenties that I started exploring the wilderness again. I was in awe of the Golden State's natural beauty and I wanted to visit these areas more extensively. I joined hiking and camping groups, such as the Sierra Club and California Adventures, where I met like-minded people. I learned how to hike, camp, and backpack in the many local, state, and national parks. I also took up rock climbing, which entailed going to places off the beaten track.

Because I was a fast hiker, I would often arrive ahead of my companions and once there hang out and watch the wildlife and nature in solitude. I grew to enjoy hiking alone, but also learned to be cautious. If I could not see far ahead, I would sing and clap, especially in an area with a lot of wildlife. My directional ability was good, so I didn't worry much about navigating my way. I always carried pepper spray and a body alarm—a rectangle-shaped plastic box with a pin that, when pulled, sounded like a car alarm. I grew more and more comfortable and was going to a local park, Tilden Regional Park, two or three times a week to hike and commune with nature.

As I progressed in my exploration of the California wilderness, I joined in overnight trips, during which I would sleep alone in my tent. This was a huge challenge. I recall a trip with the Sierra Club in Death Valley, California, where I froze most of the night because I did not have a sleeping bag designed for the cold temperatures there. Although I wore every bit of clothing I had, the cold kept waking me up. To make matters worse, a small animal kept trying to break into my tent. The animals' scratching added to my irritation. I would hit the side of the tent and it would stop. Just when I would settle back in, the scratching would start up again. Eventually, experience and safety equipment, such as a medical emergency survival kit and a Swiss Army knife, helped allay my fears of sleeping alone in the wilderness.

What started me on my photographic journey was an expensive, point-and-shoot film camera, which my then boyfriend in California, Glen, purchased for me when I was in my late twenties. I took to it immediately, and carried it everywhere I went. Adult-school classes in photography centered on black-and-white darkroom work and introduced me to the world of the single lens reflex camera. I spent hours taking pictures in natural settings and developing and printing them.

Soon I decided to make a career change and follow this passion for photography. I had been a registered nurse for about ten years before my pursuit of photography. Nursing proved to be both a practical and a healing career. As a young girl in New Jersey, I had a very close relationship with my grandmother, Mary Toth, with whom we lived. Her sudden death when I was seven left a large hole in my life. I missed her terribly, and her love and my loss led to my interest in working with the elderly. I entered a nurse's aide training program in high school, worked in nursing homes throughout my junior and senior year, and then became a registered nurse after college.

Over the years I did hospital nursing, home health care and hospice work. In the latter area, the stress of working with the terminally ill caught up with me and I ended up falling into a deep sadness. However, this dark time led me on a path of self-exploration and life-affirming activities culminating in the discovery of photography. I was ready for a change. I remained a part-time nurse to help pay the bills, as I gradually moved toward a career in photography. I worked the night shift in the hospital and took photography classes at a junior college by day. I always felt sleep deprived, but the stimulation of the classes helped me tolerate the nights. Before long I became the photo editor for the college newspaper and was employed for five years by Laney College in Oakland, California, to do their news and corporate photography. In addition, I had a wedding photography business.

Nature photography took hold of me when I began to make pilgrimages to many of the national parks. At first I went with friends and later with my husband, but gradually I made the photo trips on my own. Because I was in my mid-thirties, a young, not unattractive woman traveling alone, I found some inherent dangers in these trips. I adapted by learning to rely on my intuition to detect potential dangers, and I joined an auto club for emergency roadside assistance. After about seven years of adventures on which I had countless photographic opportunities, I developed a fine-art wildlife and nature portfolio.

Taking good photos is one thing; marketing and selling them is another. I used the matting and framing techniques I had learned in college and researched how to get my work into galleries and art shows. Through trial and error I carved out a distinctive professional niche in the field of nature and wildlife photography, and my photography is displayed and sold in galleries, in art shows, and on my website. I quit nursing altogether.

I feel honored to be able to devote myself professionally to the pursuit of my three passions: the beauty inherent in nature, the beauty of the wildlife that inhabits nature, and of course photography.

The following chapters feature nine of the most interesting and difficult kinds of wildlife I have found and photographed in North America: moose, grizzlies/brown bears, bald eagles, mountain lions, black bears, wolves, buffalo, polar bears, and wild horses. The sometimes harrowing stories of how I found and captured my images are followed by my unique photographs of the wildlife in their natural environments. The stories, with their emphasis on the behavior of the animals at the time of the photograph, enhance the images and together they create what I hope is a meaningful wildlife photo-story-book.

CHAPTER 1

Moose in the Mist

As I lay in my sleeping bag, it seemed as if I could feel every bump in the ground. I checked to see if the pebbles were also affecting my husband, but Andy was sound asleep. I knew that for me sleep would probably be impossible. My mind kept reliving the previous day's events.

We had arrived at Yellowstone National Park, my first time there, and the moose and elk immediately captivated us. They were fighting by the road in different areas of the park. It was the peak of the rutting season, an annual ritual when male animals will fight for the right to mate with the available females in heat. Large horned animals were everywhere.

We parked our car and quickly joined the other wildlife paparazzi scattered about photographing the rutting spectacle. Despite having only a 200 millimeter telephoto lens and a basic film SLR camera, I was able to get some good shots. As a community relations photographer for Laney College in Oakland, California, I had an extensive background shooting all types of news photography. My experiences taking action shots of the college's football team had made me adept at capturing the moment. But the adrenaline rush of football photography could not compare with the excitement of this new photographic experience at Yellowstone.

That night, what little sleep I finally managed to get was fitful. I kept dozing and then waking up, and I eventually decided to venture out alone in the dark to look for wildlife. I quietly dressed, packed my camera gear, and left a note for my husband. It was a fateful morning that would cement my interest in wildlife photography and lead to an almost twenty year career.

While I drove in the early morning on the road that led from the campground, out of the corner of my eye, I caught sight of a male and female moose in the river north of the Fishing Bridge area. I parked in a pullout and ran through a small patch of woods to the river's edge. The sun was just beginning to rise. The male moose was enveloped in fog, and the sunrise cast a yellow glow. It was still too dark to focus the light meter off the moose, so I went for a silhouette shot by metering off the yellow sky. Because I was using only a 200 millimeter lens, and with the moose relatively far away, I wasn't really sure what type of image I had captured.

A loud noise near my vehicle broke the spell of that moment. Afraid it could be a bear breaking into the car, I started back through the woods. Within a moment, I realized it was just another photographer who had slammed his car door. He ran down to the water's edge, going after the same shot I hoped I had successfully captured. He pulled out what I thought at the time was a massive lens and quickly started shooting. As he clicked away, he kept repeating, "That's a 'money shot.'" I assumed he was a professional photographer, and I felt a little intimidated by him and impressed with his equipment. The moose soon moved into the dense woods, and the encounter was over.

As I headed back to my car, I felt a bit deflated because of my lack of experience and "expert equipment." But to my surprise, when I later developed and printed the images, my little 200 millimeter lens and fairly basic camera had produced what the other photographer referred to as a "money shot." My "Moose in The Mist," the photo of a lone-silhouetted moose standing on the shore of a river surrounded by white mist and the golden hues of the sunrise, has remained one of the best nature photographs I have ever taken. It has sold extremely well in galleries, art shows, and websites where it has been featured; has won awards; and overall has been one of my most popular fine art, wildlife images. That fateful pre-dawn decision to get up and see what the day had to offer turned out to be the catalyst that propelled my career as a wildlife photographer. When I think back about this seminal photo experience and the moose that changed the course of my life, I hope that he and his female companion have enjoyed happy lives in the beautiful and pristine Yellowstone wilderness.

My second major moose encounter occurred on a trip alone to the Yellowstone-Grand Teton area in the mid-1990s. Since I'd decided to create a career showcasing wildlife photography, I was working on acquiring a portfolio of images that might lead to a gallery show. In addition to the "Moose in the Mist" Yellowstone photograph, I had gotten good images of heron, elk, egret, buffalo and coyotes. For this trip, my wish list included getting a more detailed close-up of a moose.

It was the first week of September, a time when the tourist season was winding down and the horned animals were getting ready for mating and winter. A lot of agitated and fighting wildlife were all around. As a beginner in the area of wildlife photography, I had acquired few tools other than my love of nature and my comfort in the woods. But I had upgraded my equipment to include a motor drive, a sturdy tripod and some large telephoto lenses, including a 500 millimeter, which was my favorite.

Alone this time without my husband, I'd had to gear myself up for the trip. In my early thirties now and fearful of who or what I would experience during my travels, I would get panicked, especially the night before I was to leave. I would toss and turn restlessly and visualize all types of serial killers. As a precaution on the road, so as to not attract attention, I always wore men's baggy shirts, older jeans and my hair tucked into a baseball cap. No makeup was also the rule. Yet as afraid and anxious as I was, I would never let that stop me from doing what I wanted to do. I always felt it was worse to be paralyzed by fear than to fail.

On the drive from California to Yellowstone, I slept in K.O.A.'s and national park campgrounds and the trip had been uneventful. The vehicle I was now using was a pickup truck with a camper shell. I carried a futon in the back and some homemade curtains, so I could sleep in the cargo area of the truck. Occasionally, I would stay in a motel for the welcome relief of a shower and a real bed, especially in the rain.

Once I reached Wyoming, I decided to go to an area in the Grand Tetons National Park that I had recently discovered. I had to access it via a dirt road that ended near a small walking bridge. From the bridge, various trails went off around a small wetland peninsula. It was early morning, so only a few people were present. Down the trail came a young couple, giddy and excited. With great enthusiasm they related that a moose was a short distance up the trail. The trail skirted the water's edge. On a spit of land, partially in the water eating vegetation was a cow moose with a small calf. It was a very serene moment and I quietly set up my camera equipment and took many pictures.

After a time I continued up the trail, where thick, prickly willow bushes grew densely. I stooped low and bent over to get through this maze of bushes. It enveloped me like a tunnel and contained a strong musky odor, scattered moose fur, and moose dung. My intuition kept telling me to turn back and go no farther. My visibility was only about ten feet, so I absolutely knew I shouldn't be there. I kept thinking, *this is crazy—you could run into a bear or moose up close and then you would be toast.* I decided finally to just go around one more corner and then turn back if nothing was there. I kept telling myself the same thing for at least four corners. Abruptly, the landscape opened up into a large area. In front of me was a lagoon surrounded by dense, tall willow bushes, and across the water was a large bull moose with a female. It was obvious they were a pair. They made tender cooing noises. It was captivating, and I set up my tripod and began photographing this intimate scene.

To my surprise, after ten minutes, out of the maze of trail from which I had just come, two large bull moose came into the clearing. They walked into the water and turned in my direction. When I had entered the lagoon area, I moved about thirty feet to the right, so that is how far away I was from them. The moose stared me down, their heads with full racks of horns tilted downward and their eyes penetrating through my body. It seemed I was being given an ultimatum: get out of here or face a charge from both of us.

Being incredibly anxious, I tried to break their stares by turning away from them, but when I turned back their stares had only grown more intense. If I could have wrapped a cape around myself and disappeared, as in a magic trick, I would have complied immediately. I silently prayed for a plan and in an instant grabbed all my equipment and started backing away slowly. But I wasn't paying attention, and I backed up to the edge of the peninsula.

Water was on all sides of me. It was either jump in and swim or try to get back to the maze trail I came on. With my equipment weighing me down and a feeling of no longer being connected to my body, I jumped into a bunch of prickly, thick, willow bushes. I panicked and thrashed about blindly. I scrambled through the massive thicket of brambles, eventually clawing my way back to the midsection of the trail, well away from the moose. With multiple scratches on my arms, face, and legs and a bruised ego, I nevertheless was happy to make my way down the trail. I kept going, not looking back, until I reached the trail end. My heart continued to pound as I quickly climbed into my vehicle and gladly left the moose "hideout."

The lesson I learned is that there are certain places you shouldn't go, and an animal's den is one of them. Although I've made many subsequent trips back to the Grand Tetons and Yellowstone area, I have never gone back to the moose "hideout." This was my first terrifying wildlife encounter, and I felt a bit traumatized. I knew I would have to be more vigilant if I wanted to survive my new career. I also became a broken record, repeating this horrendous encounter over and over when asked about my moose images. It was as if the more I talked about it, the less trauma I felt: but here I am some twenty years later relating it with a lot of the same feelings I felt back then.

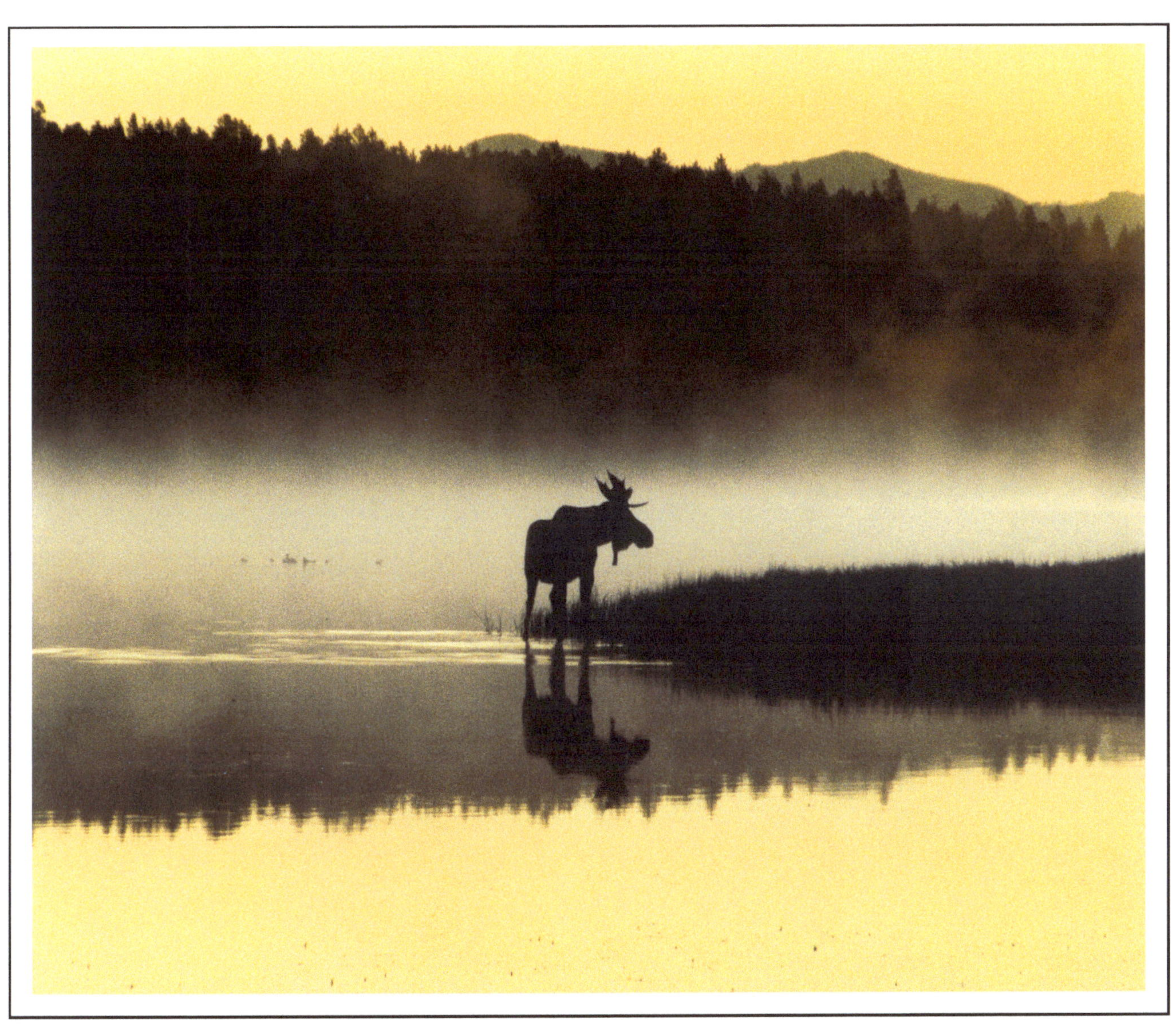

Moose in the Mist "The sun was just beginning to rise. The male moose was enveloped in fog, and the sunrise had cast a yellow glow." (page 9) Yellowstone National Park

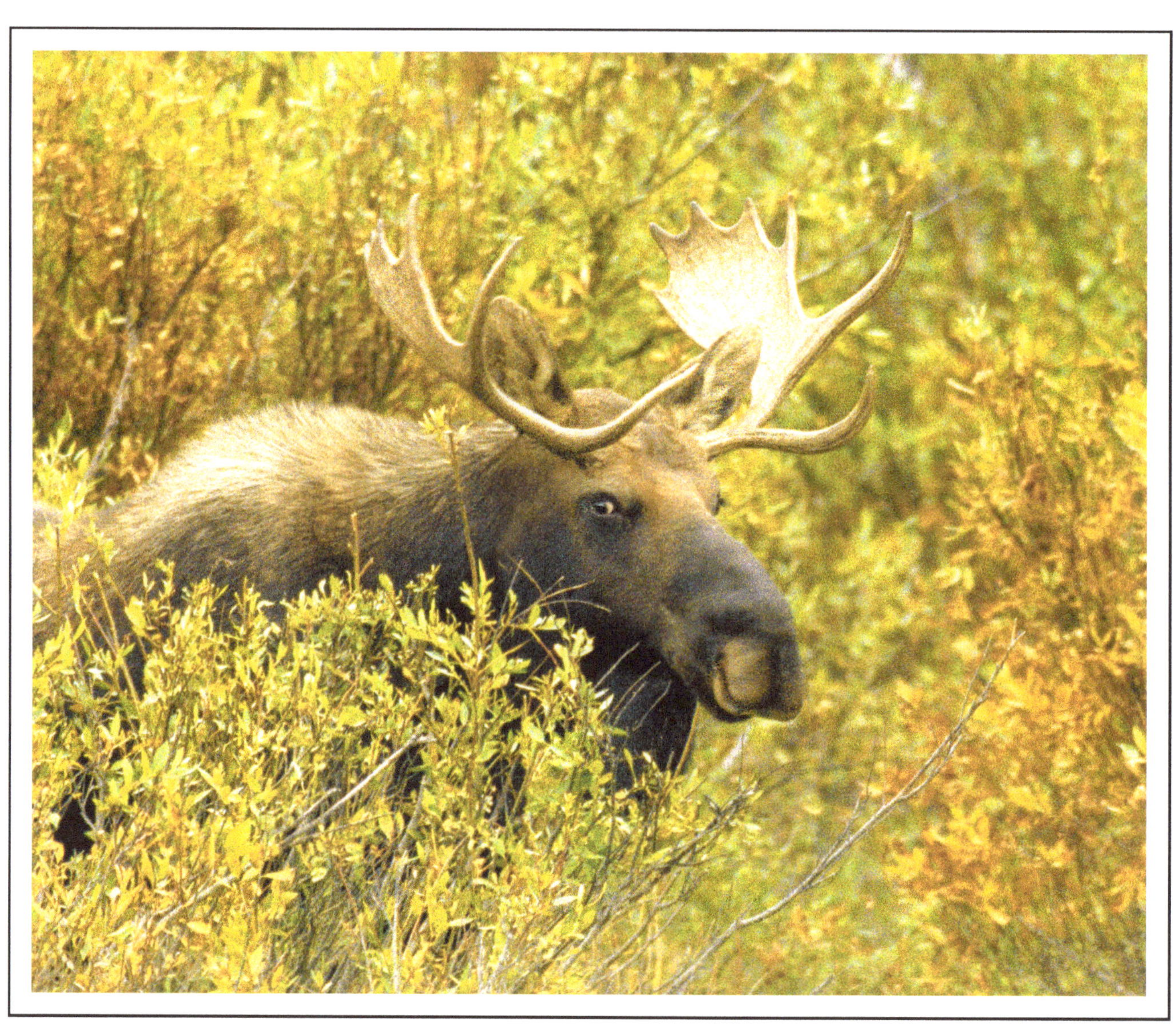

Bull Moose "Abruptly the landscape opened up into a large area. In front of me was a lagoon surrounded by dense, tall, willow bushes and across the water was a large bull moose with a female." (page 11) Grand Tetons National Park

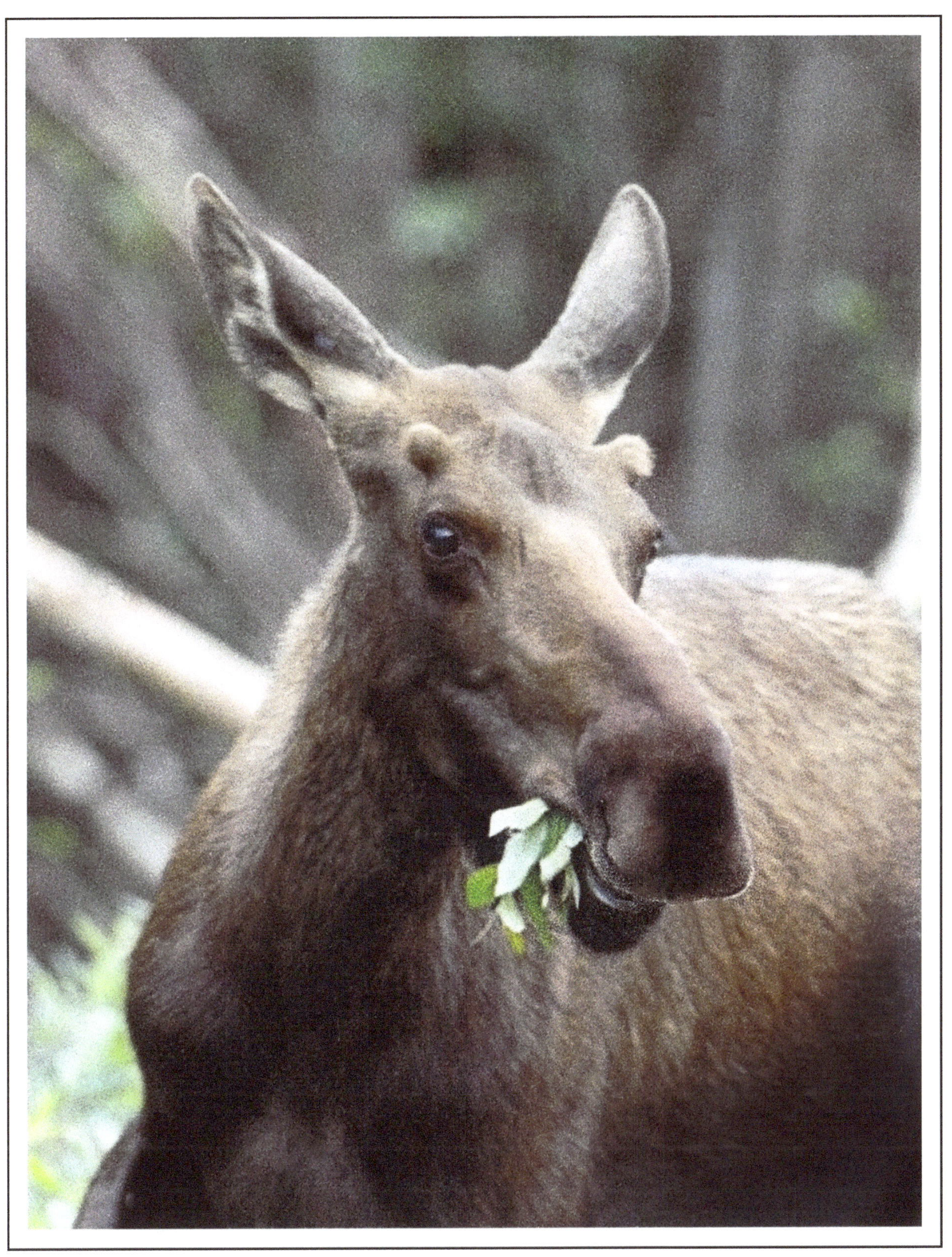

Moose Eating Grin Yellowstone National Park

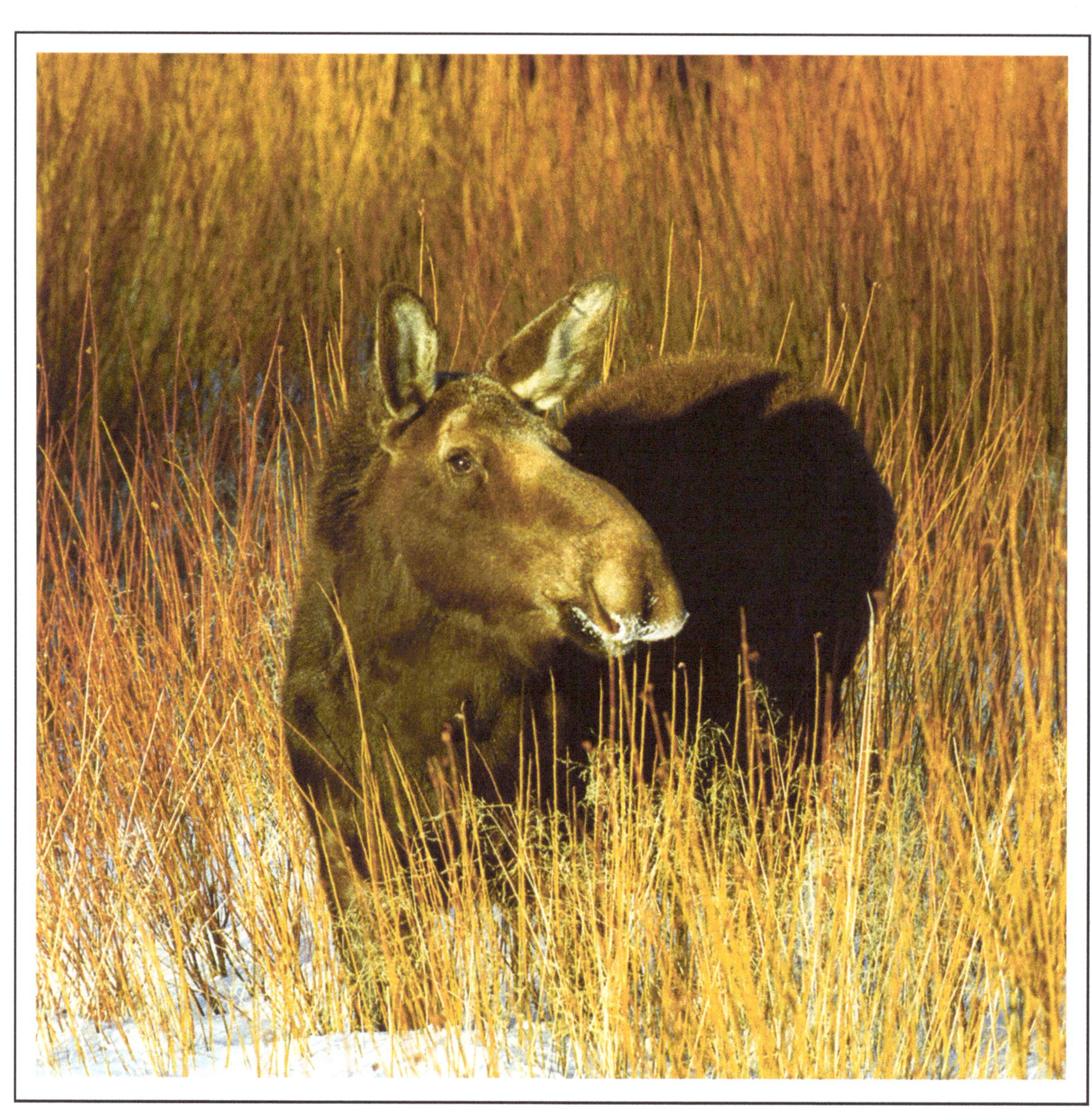

Moose in Autumn Grass Grand Tetons National Park

CHAPTER 2

Grizzly and Brown Bears

My first bear encounter with a brown bear occurred in grizzly country in Alaska. At the time, my husband, Andy, and I were walking to Brooks Falls in Katmai National Park, Alaska, during the peak of the salmon run in August 1999.

As I mentioned in the introduction, Katmai is a remote national park on the Alaskan peninsula, across from Kodiak Island. The park covers roughly more than four million acres but has no road access; it can be reached only via float-plane or boat. Brooks Falls is part of the Brooks River, which has a very prolific annual salmon run. Many grizzly bears gather at the base of the falls where salmon accumulate in their attempt to jump the seven-foot falls and return to their spawning grounds. The bears are able to eat their fill of salmon and store the protein needed for the winter.

On our first day at the park we had received our bear orientation, which included the following admonitions: If you encounter grizzly bears on the trail, don't look them in the eye. Back away slowly until you are more than 150 feet away. Always give them the right of way. And make a lot of noise while hiking. We were also instructed that there were two elevated platforms, one at the beginning of the trail to Brooks Falls and one at the end of the two-mile hike. The ranger at the departure point would tell us when we could leave the comfort and safety of the elevated platform and use our new bear-avoidance skills of yelling and clapping to let the bears know we were coming.

Wait! Isn't the ranger, a volunteer, or a docent with a gun supposed to take us out to see the five hundred-to-thousand-pound grizzly bears? Wrong! On the trails of Katmai National Park, unless you are on a guided trip, you and anyone accompanying you are alone, and you're expected to rely on your own loudest voice to yell, "Hey bear!" I immediately got very good at making noise, and even improvised by adding clapping to my bear-avoidance skills. If this was all I had, I was going to really use it. Surprisingly, I would later find out this bear-avoidance technique works really well.

The hike to the falls entailed traveling on a fire road, going around a few bends, and then proceeding down a narrow trail. The trail meandered through a dense forest that saw heavy bear use. The final portion, close to the falls, involved a hill. It felt dangerous there because there was limited visibility before and after it. As Andy and I crested the hill, I heard a loud rustling noise behind me and turned to see a grizzly bear running at us. But when the bear saw us, he veered around and kept running. It happened so fast we could only process it afterward. My husband and I looked at each other and reasoned, "Why would a bear be running—unless he was being chased by a bigger bear?"

We hurried down the trail until we came to the safety of the observation platform by the falls where the ranger with the gun was stationed. It was then that I could feel fear permeating my body. With my heart pounding as if it would rip through my chest and my mind reliving the encounter, I realized how dangerous this adventure was going to be! I could see that during my time at Katmai, I would have to struggle to maintain my composure. It made me grateful for all the times when, as a nurse, I'd had to function expertly in emergency situations. I realized that acting first and feeling later would help me to not panic.

When I looked out over the falls area, I was first struck by how enormous grizzly bears are. About fifteen were in the area, a few close to one thousand pounds. A grizzly mom and two older cubs were directly below the elevated platform. They seemed almost close enough to touch. My eyes started tearing up and a wave of warmth came over my body—I was overwhelmed by the image of these magnificent bears fishing on the river as they must have done for centuries.

It was amazing to see how bears related to each other. The two on the top of the falls had a dominant position. Perched on rocks and leaning over the falls, the bears tried to bite the salmon as they jumped up the seven foot falls. Frequently the bears missed, but occasionally they got the fish in midair. When another massive bear wanted this choice spot on the top of the falls, a kind of duel would begin between the two large bears stationed on the top of the falls and the intruder. The new bear would amble its way in between the two stationary bears and proceed to open its mouth wide and bare its massive canines. The other two bears would inspect the teeth with nervous tension and then slowly one of the bears would retreat backwards, forfeiting the number one fishing spot. I was struck by how orderly this was and how no bear was injured.

Other bears were below the falls, diving underwater, catching the fish with their paws or teeth, and bringing them up to feast. One large grizzly bear that was easily a thousand pounds was so proficient at catching fish this way the rangers gave him the title of "Diver." Another bear below the falls, fishing in a position similar to Diver's, would eat his salmon by pulling off the skin to expose the fatty meat, gobble it up, and leave the rest for the seagulls. After the meal, he looked into the sky and shook his head back and forth like a washing machine on a slow cycle, then settled down in the water with an ecstatic look on his face. He made me chuckle because I could relate to that very satiated feeling. A young blond bear caught my eye because she would whack the fish out of the water and then comically pounce on it. It struck me how humanlike the bears were—especially my comrade in eating, the food-loving brother bear.

A young mother bear walked onto the scene with a very young newborn cub, who was hanging onto her back for dear life. They stayed only a few minutes at the falls. She appeared agitated as she surveyed the scene from the shore. She snorted a few times and then quickly left with the tiny cub in tow. One of the people on the platform shared that two days earlier, one of the larger older bears had brutally killed a cub at the falls. Fortunately, I wasn't there when it happened—it would have left an indelible memory.

The ranger was strict about the times allotted to stay on the platform and asked us to leave. Another couple who were scheduled to leave five minutes later nervously asked if they could come with us. Having just seen the agitated grizzly mom leave in the direction we had to go we welcomed the company. The fear of meeting up with this agitated grizzly mom made us clap and yell, "Hey bear," as if we were all cheerleaders at a football game. Fortunately, it worked out well and we had no sightings of the bear and her cub.

This trip to Katmai ended up being cut short due to my allergic reaction to a mosquito bite. On our second day in the campground, a mosquito bite I had acquired the day before became very swollen. My face essentially grew to the size of a large pumpkin, and my eye was swollen shut. (My appearance was probably the best bear deterrent device; I imagined the bears were backing away from me!) When we arrived back at our campground mid-afternoon that day, I felt much worse. No amount of Benadryl, cortisone cream, or other drugs could stop my expanding face, and to add to my predicament, I started to have trouble breathing.

One of the air taxis agreed to take me to the King Salmon airport, where I could catch a flight to Anchorage and go to the emergency room. Sitting in the King Salmon airport and then flying to Anchorage turned out to be the hardest and longest time I have ever spent anywhere. I struggled to breathe. I kept my eye on a fixed spot, using all my effort to try not to hyperventilate. The urge to panic was very great, but I knew if I lost it, my airway might close up completely.

After I got to the ER, the physician looked down my throat and started pushing emergency drugs into my system. He told me my airway was almost closed. A short while later, I experienced a sudden blood pressure drop in reaction to one of the many medications I was being treated with. I felt myself falling backward and losing consciousness, and my body started shaking violently. My husband and I were all alone in the room until a nurse came in and dropped my head lower than my legs. This brought me around. I was required to stay in Anchorage until I recovered. My Katmai bear adventure was over.

Being so close to death shook my sense of security. For days afterward, I felt surreal, and my self-image felt very fragile. It was hard to imagine that I could go back to my old way of life.

The next part of our journey was to Denali National Park, where we camped for four days. I was terrified of getting bitten by another insect. Locals joked that the mosquito was Alaska's state bird and that Denali National Park was its primary nesting site. I became obsessed about the bugs! I took to wearing head netting, gloves, and clothing from head to toe constantly, even while I slept. I do not know what was worse: my pumpkin-like face or my hazmat appearance. Now everyone was backing away from me, including my husband.

It took an intense hike to the top of one of the smaller mountains in Denali to get reconnected with myself. The climb to the top was really difficult. We followed an animal trail through piles of loose shale rocks called scree. The climb required my complete attention because the trail was only about twelve inches wide with a steep drop downward on one side. At the top we were able to view the area around Mount McKinley with its many glaciers. Seeing the vastness around me, I had an overwhelming sense of peace and achievement. I knew at that moment I was going to be O.K.

On my second trip to Alaska, in 2005, I returned to Brooks Falls for a week. This time I was with Sandy, a fellow artist, on the trip described in the introduction. I now used a snakebite kit (the extractor) to suck out bug bites and I carried adrenaline (Epi-pen) in case of an emergency.

After our initial encounter with the female grizzly bear and her cub, Sandy and I were both shaken. And we ended up having three more harrowing experiences with the Katmai bears during the next few days.

The second encounter happened while we walked the two mile trail to the falls. Sandy and I were walking alone again when we saw a large bear coming in our direction. We immediately bowed our heads and started backing away. Without looking up, we continued to back up 150 feet, 200 feet, 300 feet—but the bear kept coming. I finally got the message and moved us off the trail. As I shuffled backwards, I glanced up and noticed the bear walking by with two other bears following. My heart pounded in my chest. One bear is scary, but three bears really shook me to my core. This was only our second day in the park and I'd already had two of the most terrifying experiences ever. We were slated to be there for four more days and my mind was spinning. How were we going to be with these bears safely?

The next encounter occurred on the Campground Trail as we walked to the lodge area one morning. We started out making our usual clapping and yelling noises. We saw a bear in the distance, so, because of our beach experience, we decided to hike through the woods above the trail. With no trail in the woods, we had to bushwhack our way through. We encountered many areas of flattened tall grasses, a musty odor, and bear scat scattered about. It reminded me of my "moose hideout" experience, and my intuition was screaming, "Get out!"

Sandy and I looked at each other and realized we were in the bears' bedroom—oops! From the look of it, it was a very well-used and popular bedroom. We also realized that a slumber party could be going on right then, and we were crashing it. The trail with the bear on it was looking better and better. We hurried back to the trail and saw that the bear was gone. So we continued on, yelling and clapping.

We were walking near a park ranger storage box about the size of a washing machine when out of nowhere a bear popped up from behind the box and really surprised and frightened us! Recoiling, we kept a large distance between us and the bear and proceeded slowly backwards without looking up. The whole time my mind was talking myself out of running, because that is what I wanted to do: run back to the electrified fence! Later, joking about it, we wondered if the bear surprised us on purpose. We could see the lively discussion taking place in the bear bedroom: "Ha! You should have seen the tourists I scared today!"

The final encounter was one of my worst wildlife encounters. It occurred on the Brooks Falls trail returning from the falls area. Sandy and I saw a really large bear coming at us on the trail. Having learned our lesson, we immediately got off the trail and started backing away from the bear slowly with our heads bowed. For some reason, Sandy froze about twenty feet from the trail and was staring at the bear. He, in turn, was standing sideways staring at her.

I was in a state of complete panic. I knew that the bear viewed Sandy's freezing and staring as a threat. Talking in a low voice, I tried to coax her to bow her head and move backwards. This lasted about five minutes, but no amount of coaxing from me could get her to move. Finally, the bear abruptly turned and moved on. It was the longest five minutes in the history of minutes.

The feeling that, the worst thing was about to happen, stayed with me for a long time after that experience. Incredibly, Sandy said, she didn't feel threatened, but this encounter resulted in many heated discussions about bear etiquette. The biggest lesson I learned from all of this was to slow down and make more noise. That was my successful strategy for the rest of the trip. I left those bears with their ears ringing!

Since my last trip to Katmai National Park, the viewing platform had been extended to remove the trail's hill section, which had been so terrifying on that trip. This was a great relief to me and allowed a much larger area for observing the bears. I became fascinated by a grizzly mom with her two cubs, standing in the middle of the river. The cubs appeared to be in their second year, and their mom was teaching them how to fish. The mother caught the dead fish as they flowed down river by going underwater and grabbing the fish parts with her paws. Her two cubs watched her intently. The larger of the two cubs followed the mom's lead with occasional success, but the smaller one would try, lose his balance, and struggle just to stay in the water. He was not the star! He tried again and again and finally sat on a rock, apparently tired, frustrated, and defeated. Eventually he abandoned the lesson and struggled through the rushing river back to shore. His expressions of frustration and disappointment endeared him to me; I focused quite a bit of my camera work on him.

On the same trip, I was startled to see two large bears, a male and a female, splashing and playing in the water. At first I thought they were fighting, but after a while I could see that they were playing. To see these large bears (both well over seven hundred pounds) interacting was amazing. The huge splash when they fell into the water together created a small tidal wave. Their expressions were so exaggerated, it made me chuckle. One would circle the other then jump on its back; the submissive one would then shake him off. This continued for an hour until finally they waddled off into the deep woods. Their playing was a mating ritual that I was fortunate enough to witness.

Another time I saw two bears fight. It lasted less than two minutes, but the power and ferocity of the bears was terrifying. A younger bear had tried to move into a territory where a large bear was fishing. The large bear was not in a sharing mood; he growled loudly, opened his mouth, and viciously snapped at the young bear. The young bear shot backwards quickly, avoiding the large bear's savage jaws. As he swiftly ran away from the area the young bear's head was bowed very low, and he was whimpering. The larger bear resumed his fishing unperturbed.

In the same area of the falls, a young female bear with two cubs tried to fish while also keeping her cubs safe. She chased the cubs up a large tree and returned to her fishing. Left alone, the cubs soon climbed down and played in the grass at the front of the tree. They rolled around, fought over twigs, and climbed on each other, oblivious to the surrounding danger. After seeing this, the mother dragged herself out of the water, barked, growled and chased them back up the tree. This routine occurred several times. Finally she smacked the tree, as well as the last cub on its way up. The cubs seemed to get the message and this time stayed in the tree until the mom was finished.

My times at Katmai National Park have been unique and I view the park as a global treasure. There are no other places I know of where wild grizzly bears have adapted to humans in such close proximity. The specialness of this relationship is something I hold dear and although I've had many harrowing encounters, I would camp with these bears again.

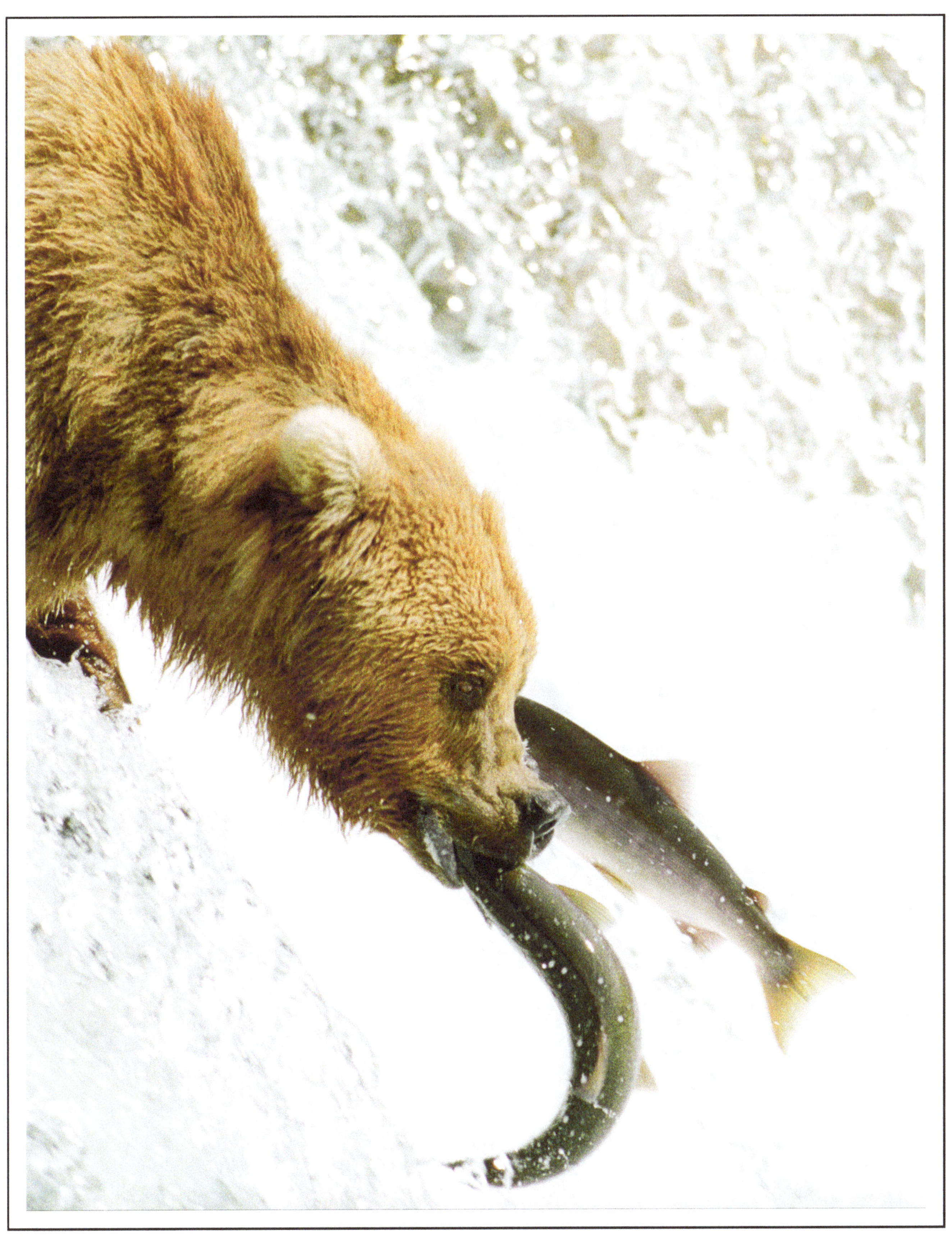

Catch of the Day Brook's Falls, Alaska

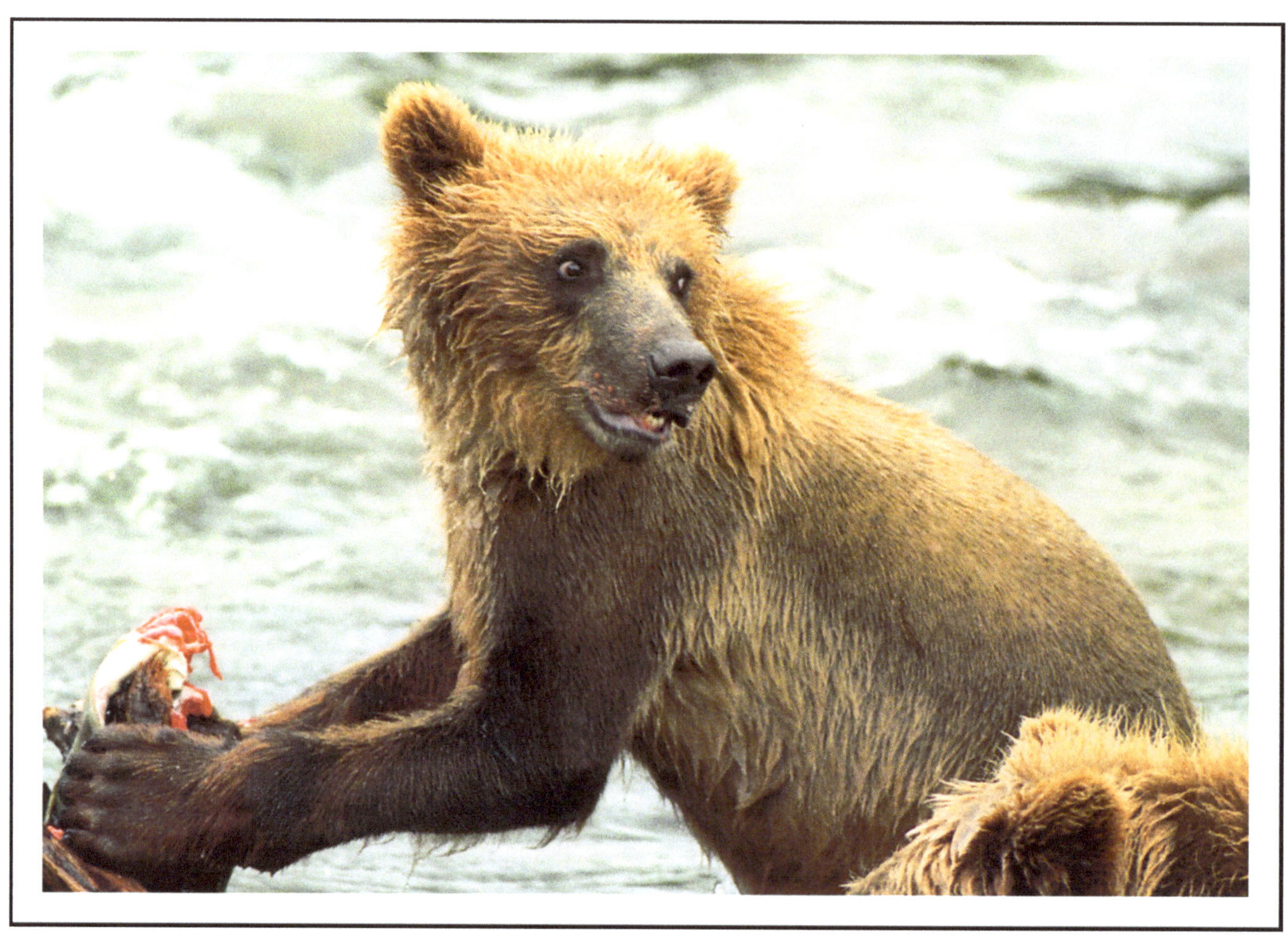

The Star Cub "The larger of the two cubs followed what she did with occasional success, but the smaller one would try it, lose his balance and struggle just to stay in the water. He was not the star!" (page 21) Brook's Falls, Alaska.

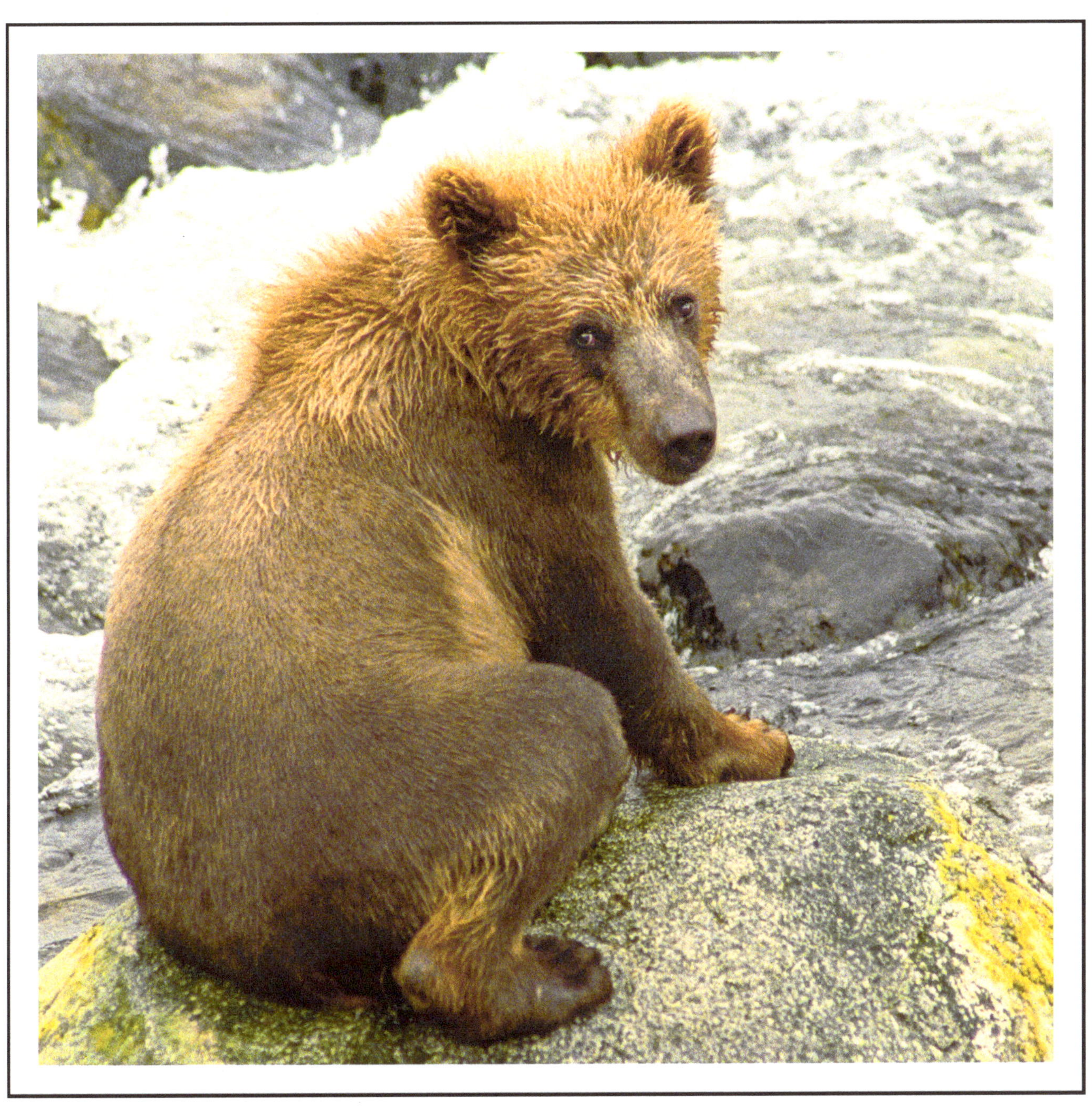

Humbled Cub "He tried again and again and finally sat on a rock apparently tired, frustrated and defeated." (pg 21)

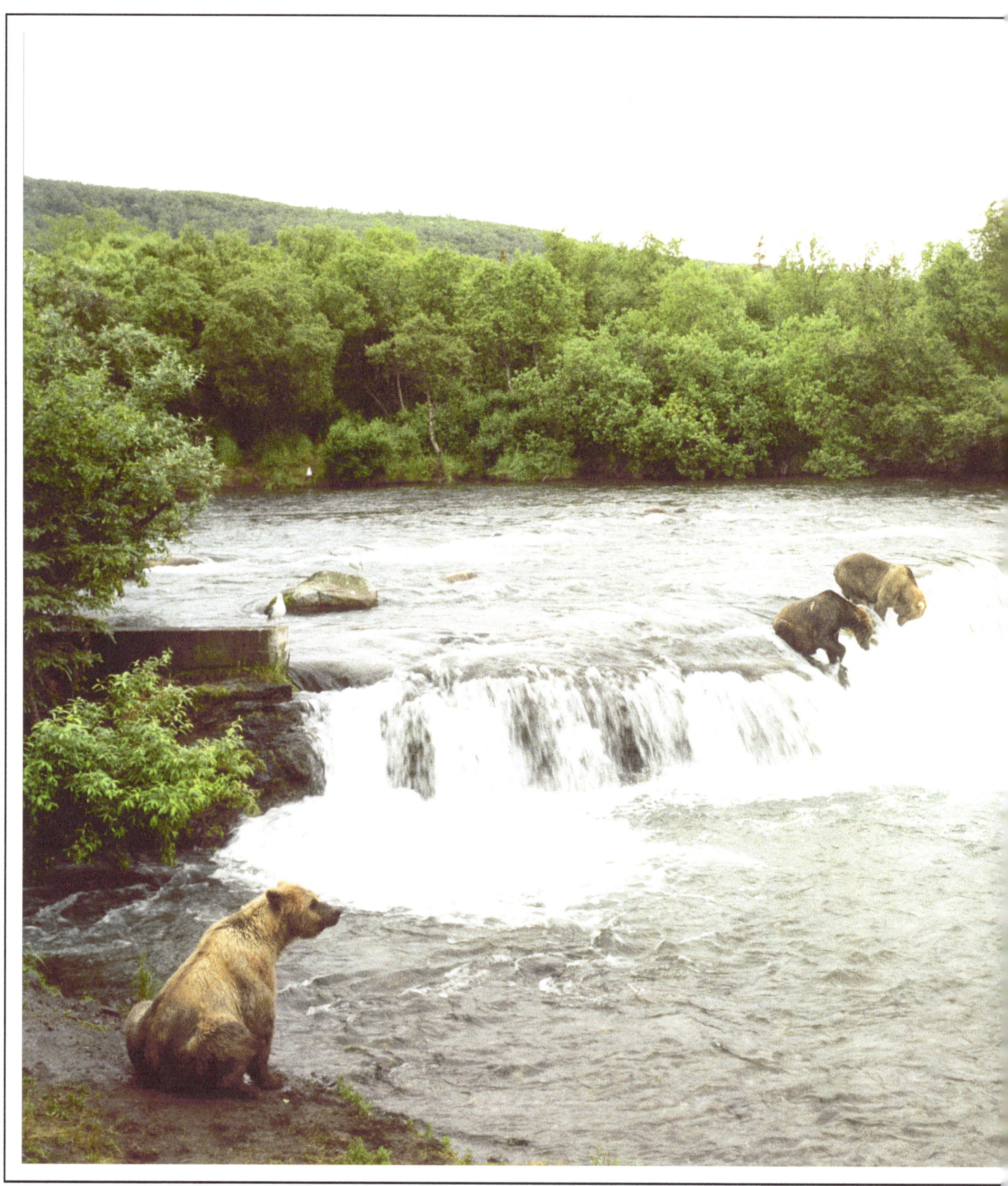

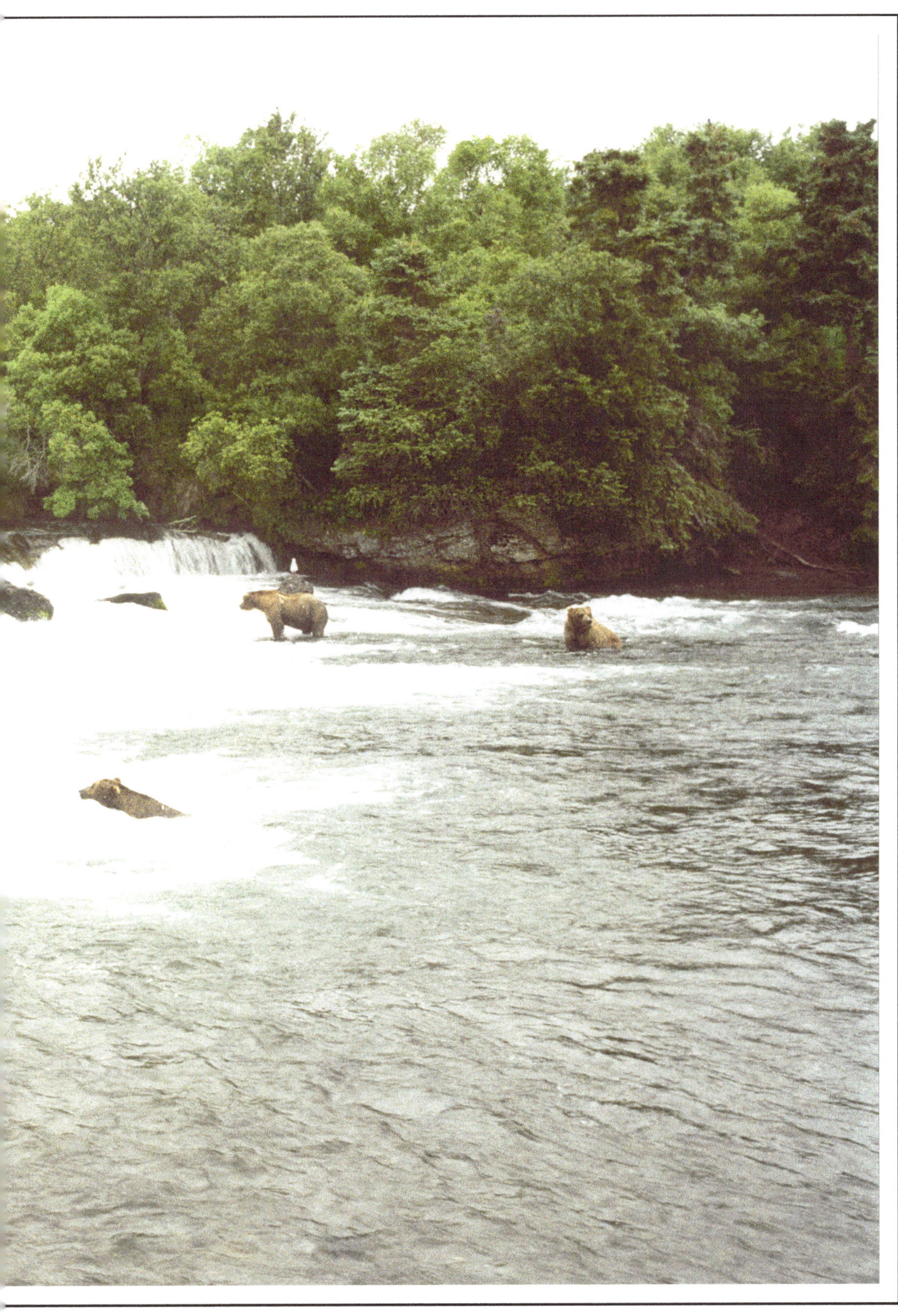

Fishing at Brook's Fall "When I looked out over the falls area, I was first struck by how enormous grizzly bears were." (page 18) Katmai National Park, Alaska (limited edition print)

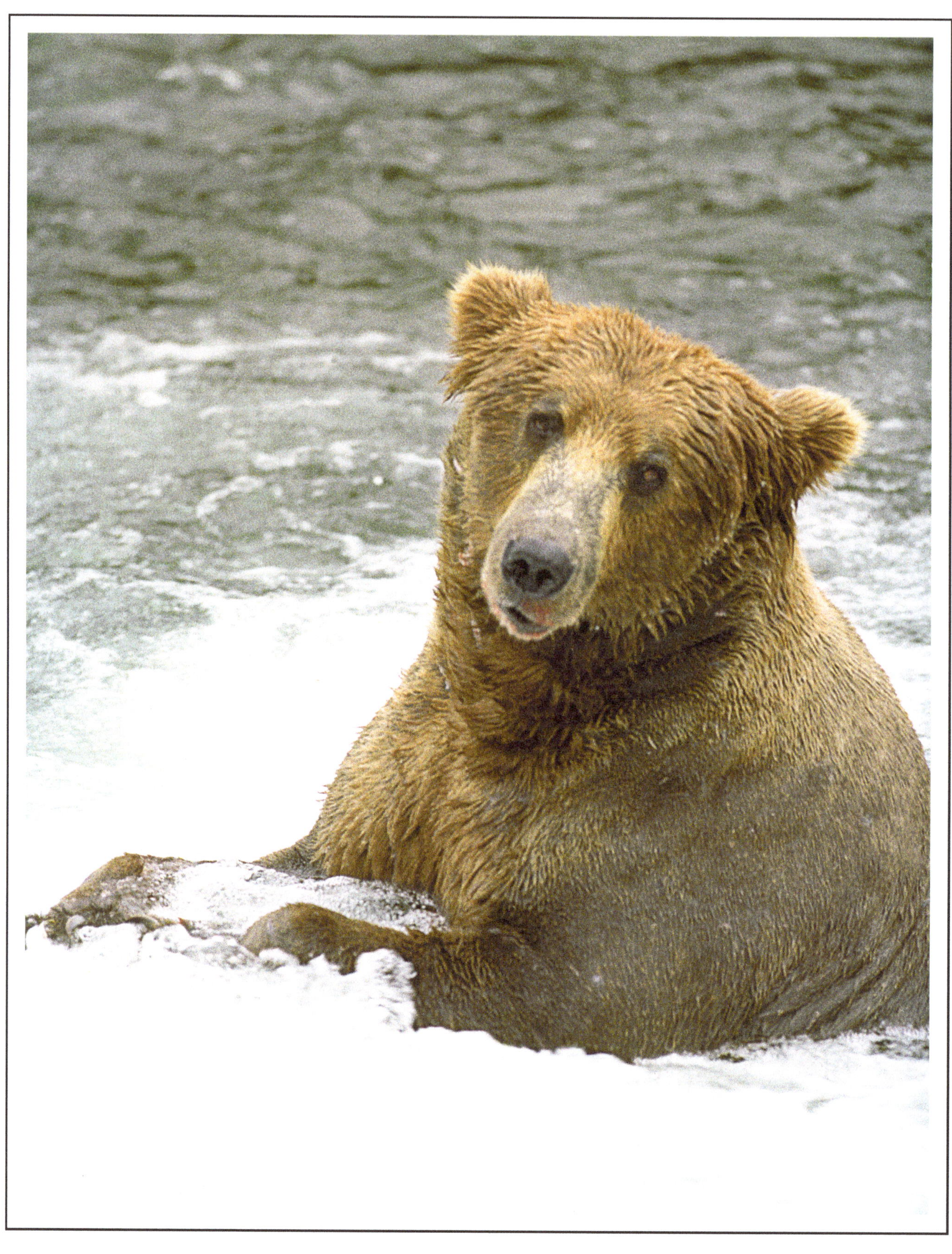

Gratified Grizzly "After the meal, he looked into the sky and shook his head back and forth like a washing machine on a slow cycle, then settled down in the water with an ecstatic look on his face." (page 18) Katmai National Park

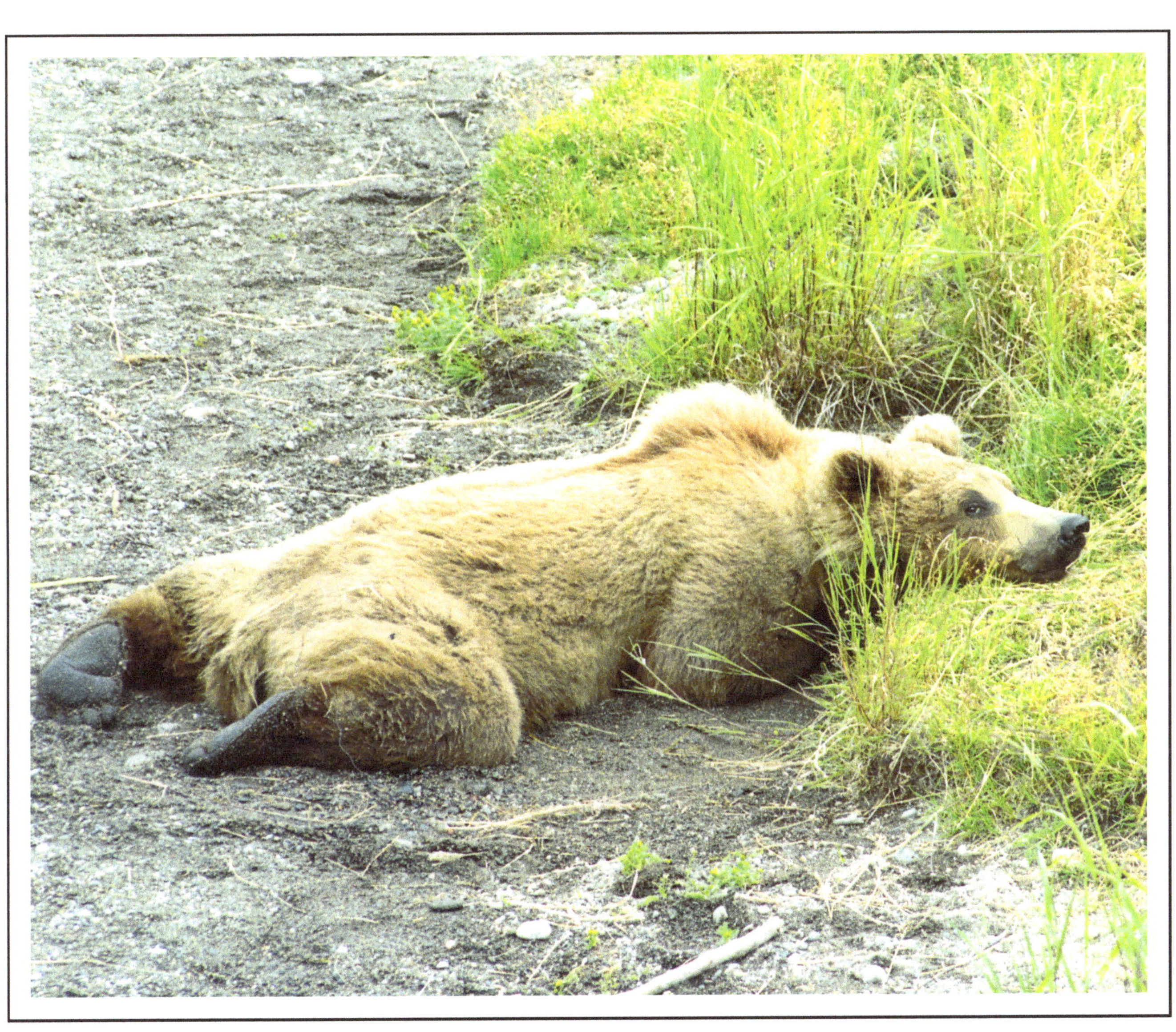

Grizzly Morning Katmai National Park

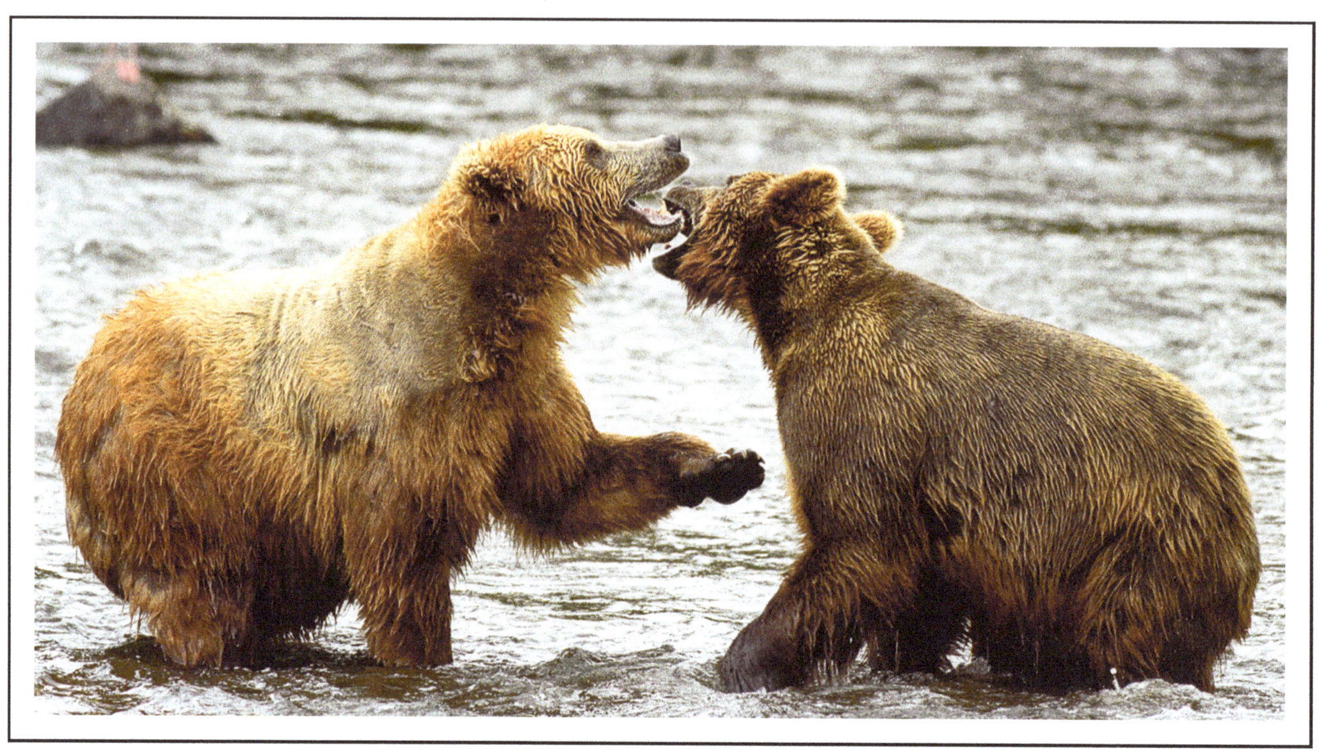

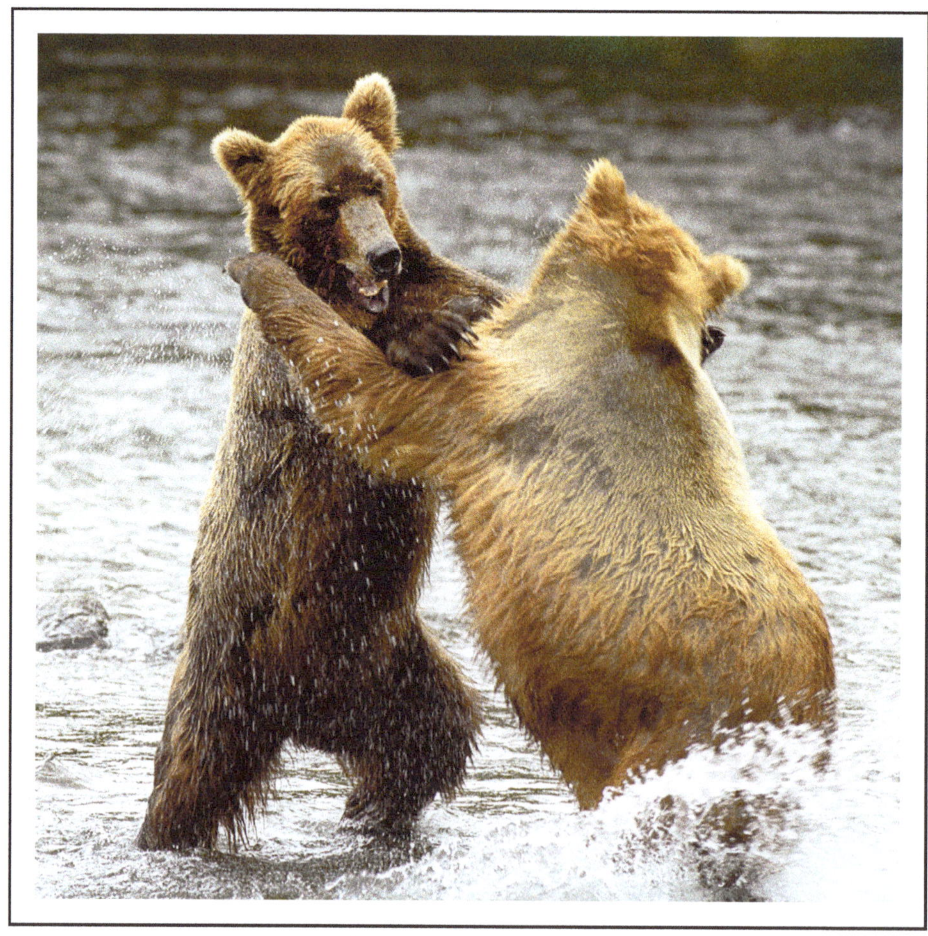

TOP: *Mating Dance* (image one) Katmai National Park
BOTTOM: *Mating Dance* (image two) Katmai National Park

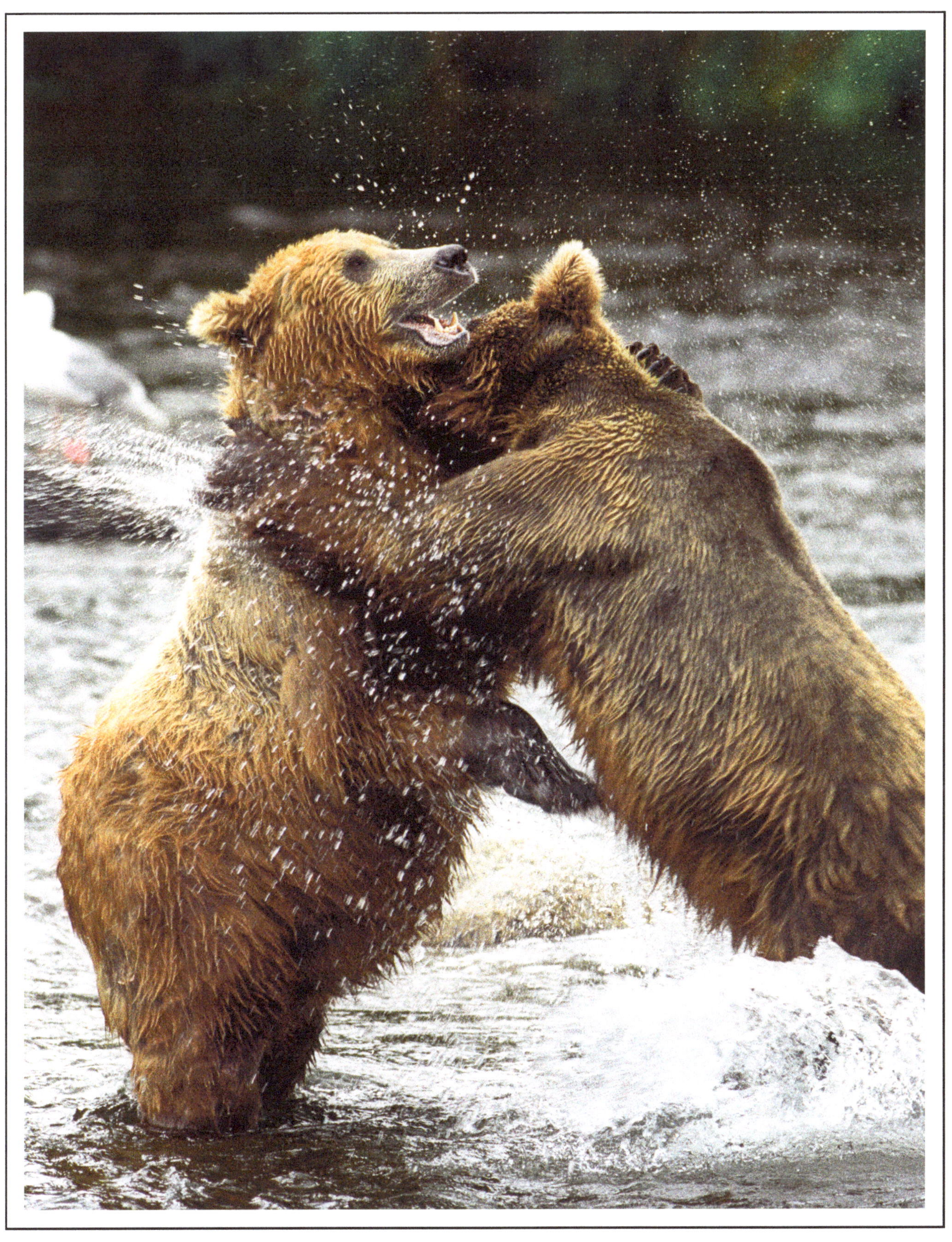

Mating Dance (image three) "One would circle the other then jump on its back; the submissive one would then shake him off. This continued for an hour until finally, they waddled off into the deep woods." (page 18) Katmai National Park

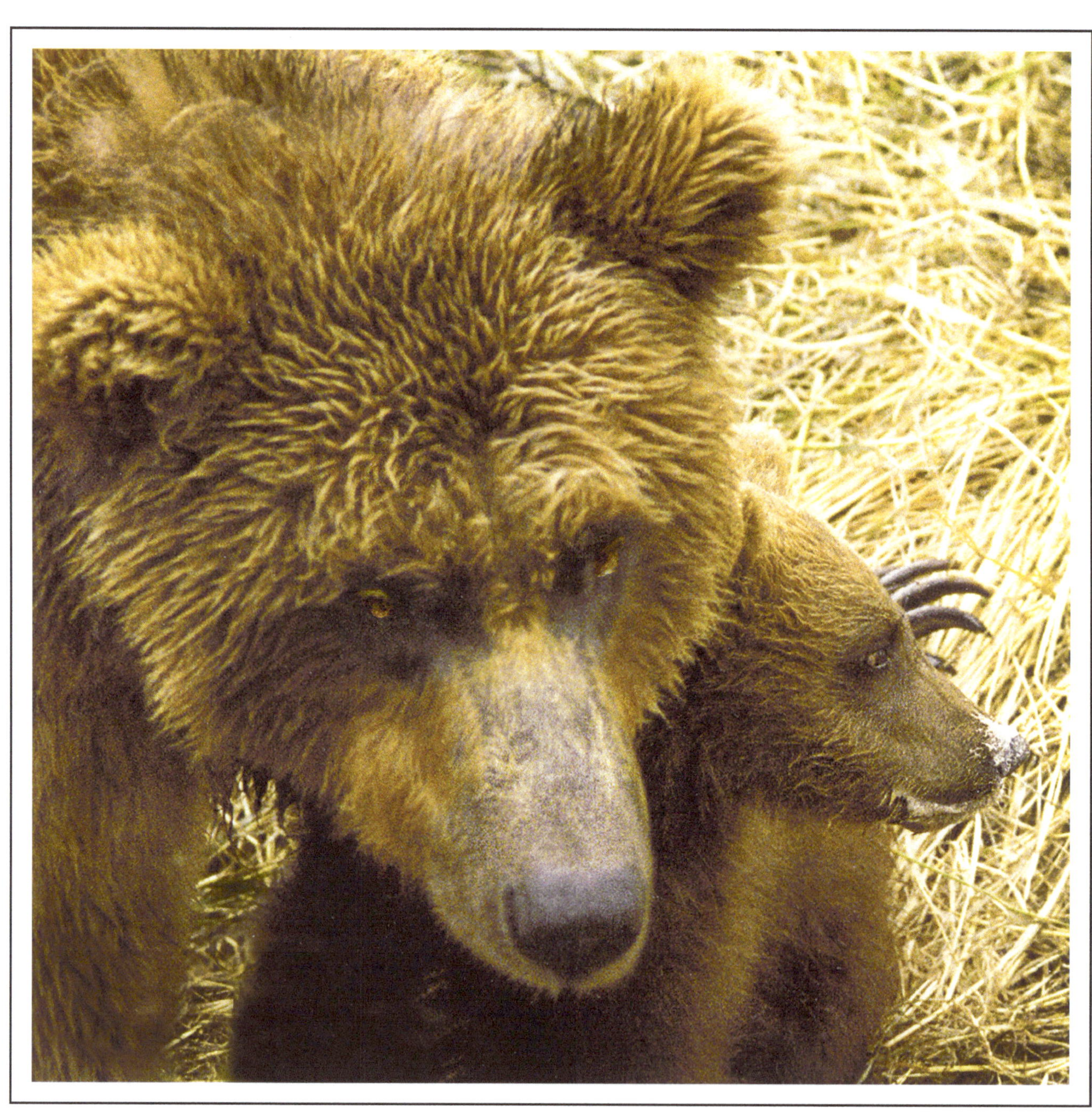

Doting Grizzly Mom Katmai National Park

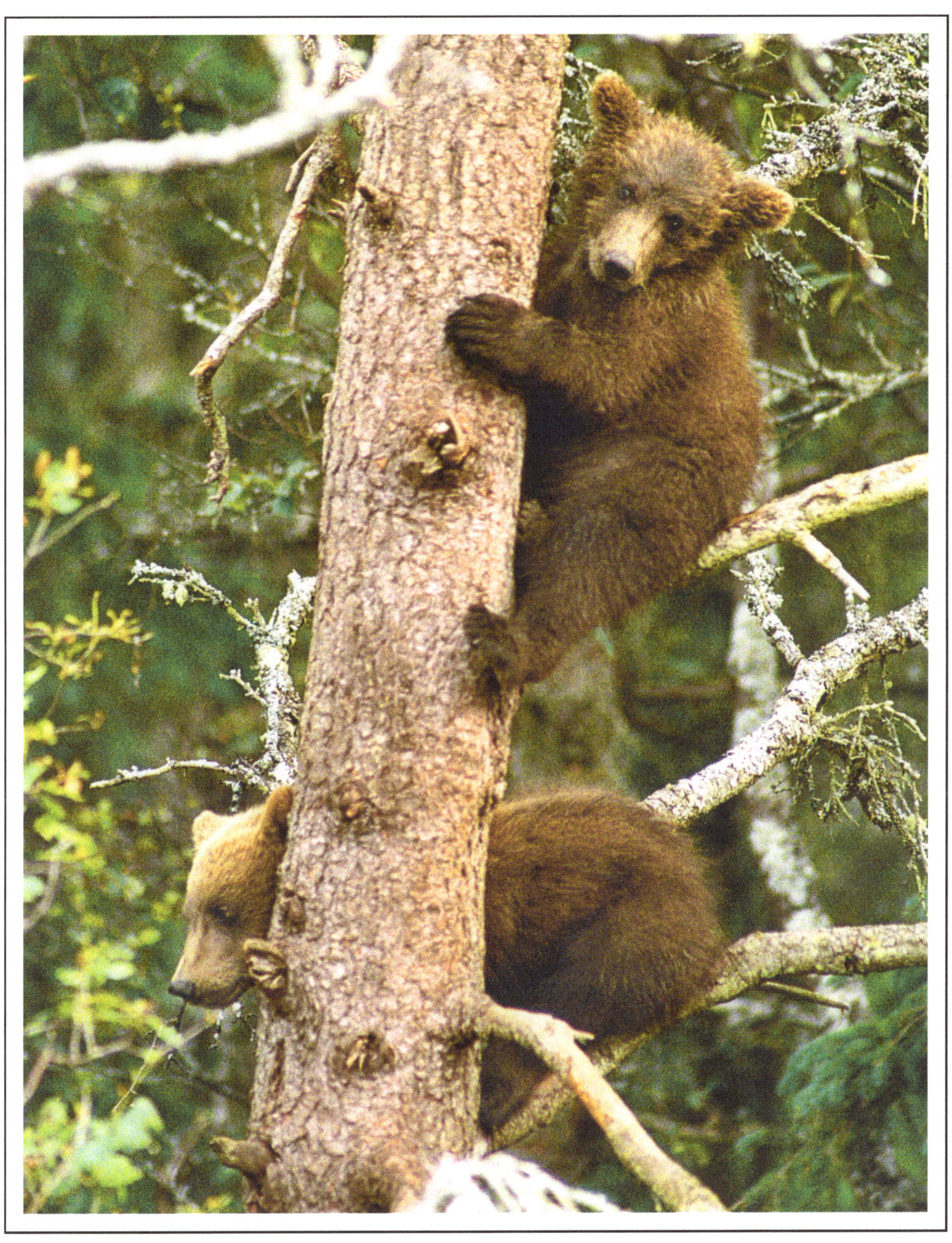

Cub Brothers "She chased the cubs up a large tree and returned to her fishing. Left alone the cubs soon climbed down and played in the grass at the front of the tree." (page 22) Katmai National Park

CHAPTER 3

Bald Eagles in Flight

Barely awake, I could feel the chill air of September in the Rocky Mountains. I forced myself out of my toasty-warm sleeping bag as the darkness receded and the day began to dawn. Dressed and with my camera gear ready to go, I sprinted to my vehicle, then drove slowly around the park looking for wildlife. It was a routine I had practiced countless times. My stalking ground this time was the Grand Tetons National Park, so named for the pointed granite peaks that rise from the surrounding valley. Located just below Yellowstone National Park and encompassing part of the Jackson Hole Valley, the Grand Tetons area includes numerous small lakes, rivers and wetlands—excellent places for spotting wildlife.

September was a time when larger horned animals, such as elk and moose, fought with each other in preparation for mating. They were the focus of my trip, but as it turned out this became the setting for my first experience with bald eagles. As I hunted for wildlife, I set out for a familiar area where I had seen moose before. I was on a small dirt road that ended at a fishing area (the area I referred to in chapter 1, "Moose in the Mist"). Halfway to the fishing area, I rounded a bend in the road and, out of the corner of my eye, noticed a bald eagle high up a tree on the other side of the Snake River.

I looked for a place to position myself, so the eagle wouldn't see me. On my side of the river was a large tree that would be a perfect blind, but to get to it, I would have to navigate an open field dotted with small bushes and tall grass. (I try to use blinds whenever possible so as not to disturb the wildlife. Blinds can be anything from a tree to a large rock or, frequently, a car.) I inched my way closer across the open field and got within ten feet of the tree. Crouched down, I pulled out my lens and at the same time the eagle's head swiveled in my direction and he quickly flew off. I didn't even get off one shot. I shrugged and thought—*Another time.*

The next morning I followed the same route and there the eagle was in the exact same spot. This time I decided to practice more stealth. I put on some "camo"—my black jacket with hood and black pants, and slipped out the other side of the vehicle. Bent over, I crawled toward the tree. When the eagle turned the other way, I rushed between bushes. At the tree, I held my breath and slipped my lens around its trunk until I focused on the eagle. As if on cue, the eagle quickly flew off. Again, he was gone before I got off a single shot. So much for my stealth behavior. I realized it's impossible to sneak up on an eagle.

I tried a different tactic when I went to Washington State. It was during the annual salmon run on the Skagit River, when thousands of salmon from the Pacific Ocean navigate the Puget Sound and swim up the Skagit to spawn before they die. The prospect of all that protein attracts as many as six hundred to eight hundred bald eagles to the river each year, the largest number of wintering eagles in the continental United States. They arrive around late October and stay well into February with the highest number usually seen in January. I went to the Skagit River in February 2004 at the height of the salmon and eagle season. After my failed attempts at sneaking up on bald eagles in the Grand Tetons, I decided on a new tactic: I would float up next to them on an inflatable raft.

I had arranged for a Skagit River raft trip in advance and was at the meeting spot when the bus pulled in. It was an old, tattered school bus with an interesting looking driver, "a mountain man," is the best way I could describe him. In his thirties, he had a muscular build, a foot-long beard that was a wild and unruly reddish-tan, and hair the same color and length. He gave me a hearty welcome as he opened the handle that swung open the doors. I stepped into the bus somewhat reluctantly—its interior was as dilapidated as the exterior. Inside was old bedding, a mattress, an assortment of soda and water bottles, food wrappers, life preservers, and dry bags to help keep equipment dry on the water. Stretched over a threadbare, three-seat-bench was an old, slow moving shaggy dog whose eyes were hidden by bangs, but who also seemed friendly enough. I looked around for a place to sit. Most of the other seats were missing. Not wanting to disturb the dog, I sat on a big metal box bolted to the side of the bus.

As we drove to the portage area, I thought, *What did I get myself into?* The driver seemed really nice, but his physical appearance and the bus sent up red flags. My anxiety lessened when he picked up a family of three who were also booked on the excursion. Two of the three were grown men, a father and adult son—and the third was the son's mother. *Safety in numbers*, I thought. Plus, they seemed a lot less bothered by the situation than I was.

We got to the portage area and saw we were going out on a midsize river raft. Portage areas are pullouts along the river without vegetation, which gives easy access to rafts going in and out of the river. This one was below a two-lane bridge and had a small parking lot. Other rafters and guides were there, getting ready for trips on the river.

I had brought along my tripod and a very heavy manual-focus 500-millimeter telephoto lens, as well as two cameras and other telephoto lenses. I had a lot of trepidation about bringing them on the raft. My first concern was getting them wet and then how was I going to use a tripod on a rubber floor? The river was also rushing rapidly, and I needed to have the raft still to manually focus my large lens. Jack, our driver, who also turned out to be our river guide, encouraged me to bring the equipment along. He was confident he could stop the raft on the raging river and excitedly told me he would take me to a place with a lot of eagles. His confident and jubilant manner sold me on his abilities and river knowledge. Also, I really needed my long lenses to get a close-up sellable shot. Wavering at first, and then relinquishing my resistance, I brought it all.

As we floated downstream, I saw eagles scattered in trees along the sides of the river. I set up my large lens and tripod, its three legs stabbing into the bottom of the raft. I kept a tight grip on the lens, afraid it would tip over into the swirling three-inch puddle of brackish water on the bottom of the raft. I thought, *One false move and your thousands of dollars in equipment will be toast, soggy toast.* Keeping a death grip on my big lens, I used my free hand and my second camera and shorter lenses to take pictures of the eagles flying close by. We came around a bend on the river where the river bifurcated. In a tree there was an eagle.

The current was horrendous. Jack quickly announced, "Be ready!" He took the boat in between the rushing split current and sidled up close to the eagle. I had gotten to know the family well as we floated along. The young man and his mother sat on either side of me, and they graciously moved their feet to the side of my tripod's legs to steady them. I manually focused on the eagle as it started to fly away. But this time, I was able to get shots, in rapid order, of all the stages of flight, from takeoff to a full frame of it soaring wing tip to wing tip.

Jack had done it! With a herculean effort, he had held the raft completely still. Mountain Jack turned out to be one of the best guides I have ever had and I thanked him profusely. I will always be grateful to Jack and to the family who helped hold my tripod. What is the old saying, "You can't judge a book by its cover"? Well you also can't judge a guide by his school bus!

After my positive raft experience on the Skagit River in Washington, I decided to try this approach on the Russian River in Alaska. The Russian River runs westward toward the ocean through the Kenai Peninsula. It is world renowned for its fly fishing and it has a very prolific salmon run that attracts bald eagles. It was the year 2005 and I was with my artist friend Sandy.

We arrived at the portage spot with all my camera gear, this time ready for anything. I was relieved to see a much larger and therefore more stable raft, than the one that took me down the Skagit River. We also had an athletic river guide who seemed very skilled in paddling. Armed with my new tripod-stabilizing skills, I was ready to go.

We drifted along the lush Russian River and saw a lot of scattered wildlife, including wading birds and a beaver. However after about fifteen minutes, we went around a sharp bend and came upon thirty-five to forty fly-fishermen, all crowded together, casting their fly lines as they tried to catch salmon making their way up the river. Our guide called them "combat fishermen." How they didn't snag each other or freeze in the ice cold river I'll never know. With all the commotion, it was an amazing sight, like Grand Central Station in New York but on the Russian River. The fishermen seemed pleased to see us, a distraction from the relaxing, yet serious task at hand, and waved to us as we drifted by.

A half mile ahead, we went around another bend and saw a large grizzly bear swimming about twenty feet off to our right, heading for shore. A little farther down was a black bear cub up a tree. No doubt there was a female black bear not far away. I was in a frenzy photographing these bears when I noticed yet another couple of large brown bears on the shoreline. All totaled, at least five bears were visible. This was a formidable group of bears. I wondered if the two groups ever intersected. As I thought about this, the river guide shared that sometimes small cubs stole backpacks left on the river bank while the owners were fishing.

As we continued down the river, I spotted my designated prey, a bald eagle on a branch. The branch was connected to a tree on a small dirt-mound island. The left side of the branch was connected to a bunch of brambles on the other side of the river. A small tributary ran between the dirt mound and the river bank, and dense vegetation grew overhead creating a tunnel-like effect. The eagle sat on the branch in the overhead bramble at the top of the tunnel.

As our raft sailed under the branch, I could not believe my luck: the eagle stayed motionless and followed me with his eyes. The curiosity of the young eagle with its engaged look sealed the raft idea in my head.

One last meaningful encounter with bald eagles reminded me that wildlife photography was rewarding not only for its technical difficulty but also for its spiritual renewal. In 2004 I had a debilitating illness that followed neck surgery for an accidental injury. For over a year I suffered significant weight loss and chronic pain. Eventually I recovered physically, but mentally I lost interest in life and moved about robotically. With encouragement from family and friends, I took a wildlife photo trip to Squamish, British Columbia to witness the annual salmon run which attracts over a thousand bald eagles.

In keeping with my favored bald eagle-stalking method, I took a raft excursion on a river that ran through a Native American Reservation. While floating through a narrow canyon enveloped in mist, a large bald eagle appeared overhead. The beauty of it so close-up and the sound of its flapping wings as it flew by permeated my soul and ignited a long-dead spark.

When the raft trip ended I asked an employee of the reservation if I could hike back into the canyon and he graciously gave me his blessing. I hiked two miles following the river bank to the canyon. I partially climbed up the mountainside and sat down. Watching and listening as the first bald eagle flew through, a tear formed on my face and before I knew it I was sobbing uncontrollably. With each eagle that flew through I was overcome with a new wave of emotion—I realized my soul was being mended one bald eagle at a time. This lasted a long time until it got late and I walked out of the canyon feeling renewed. Even today, I still get chills when I hear and see an eagle fly close by.

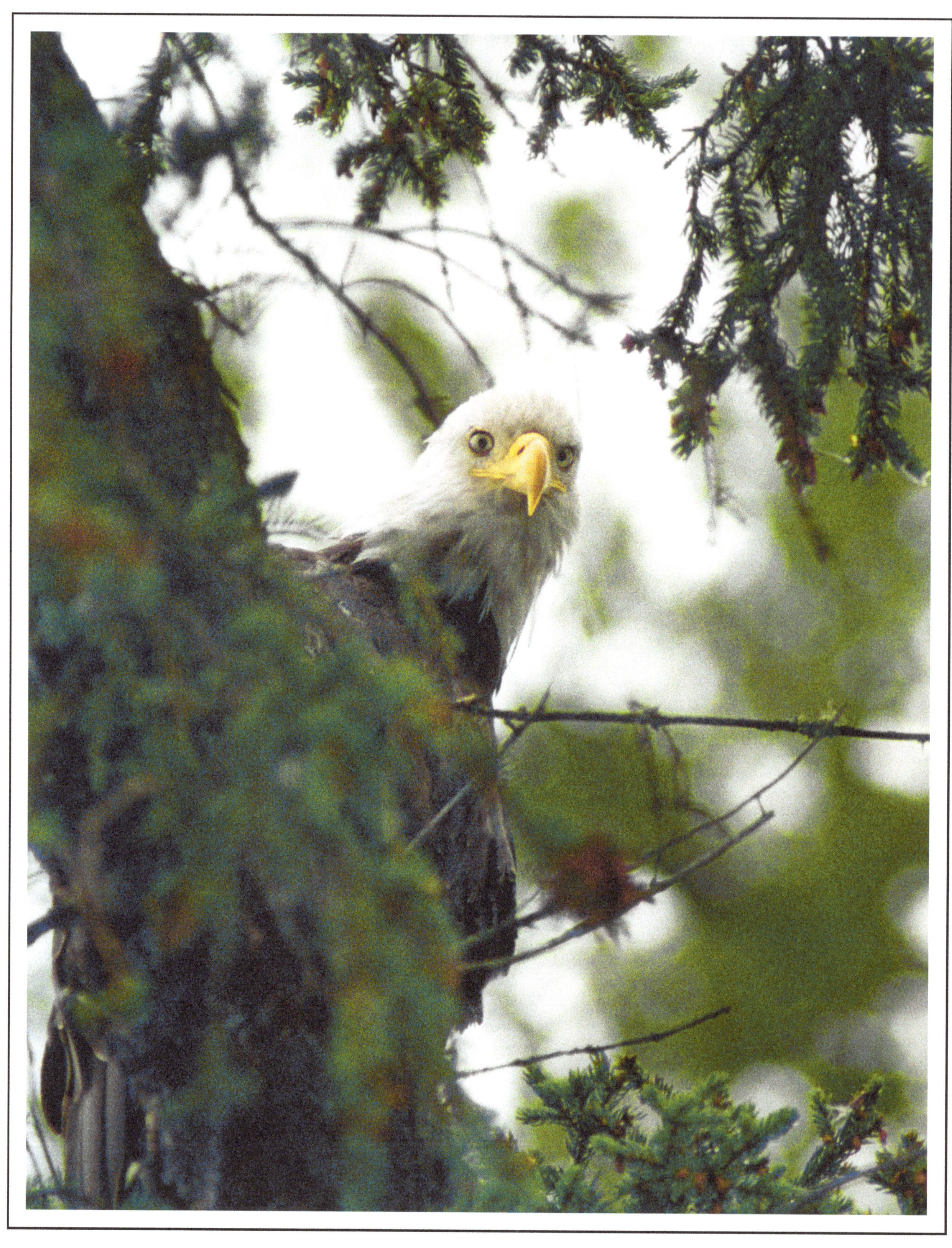

Eagle Eyes "As our raft sailed under the branch, I could not believe my luck: the eagle stayed motionless and followed me with his eyes." (page 37) Skagit River, Washington

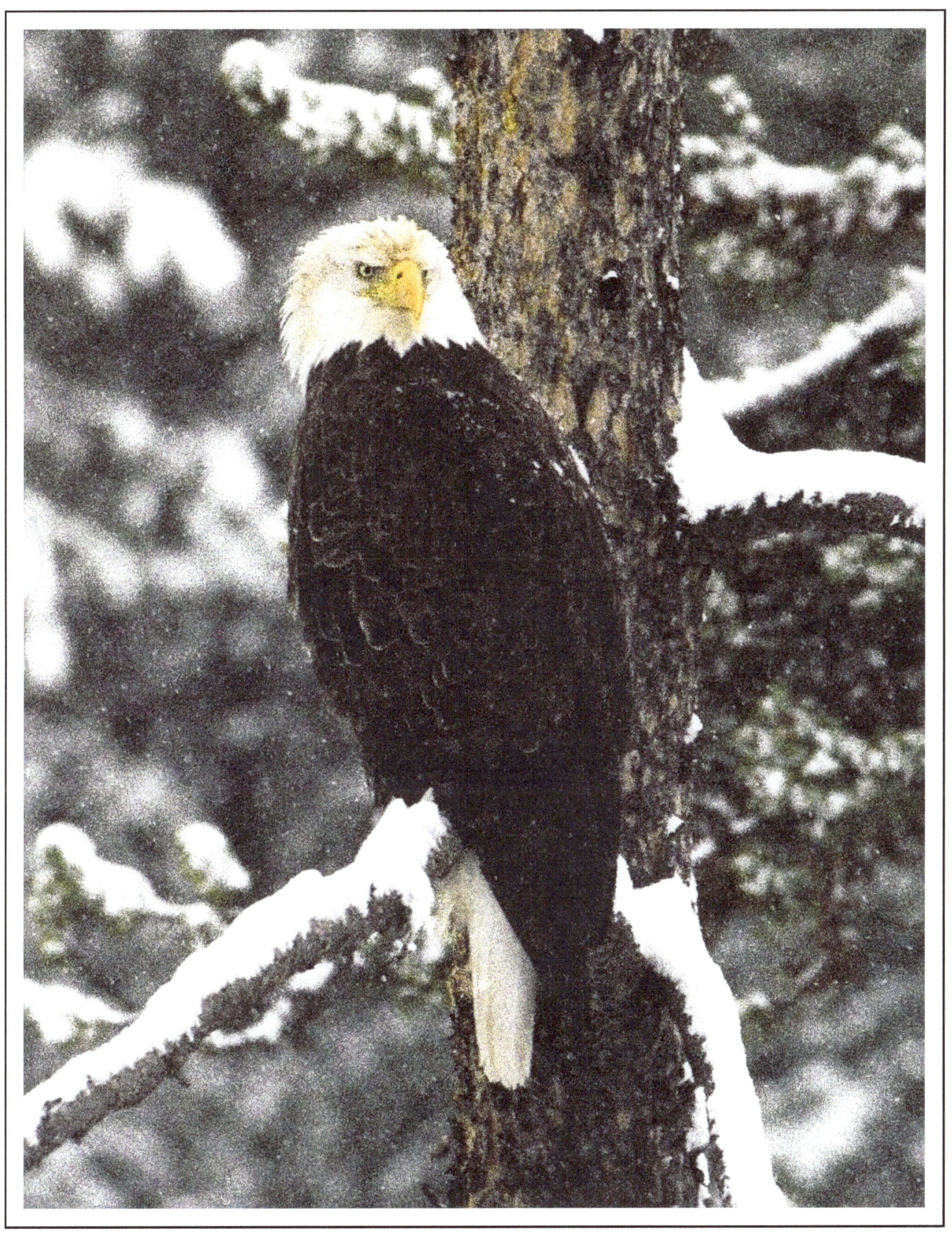

Wintering Eagle "Further down the route a bald eagle in its winter plumage sat statue-like on a branch." (page 68) Yellowstone National Park

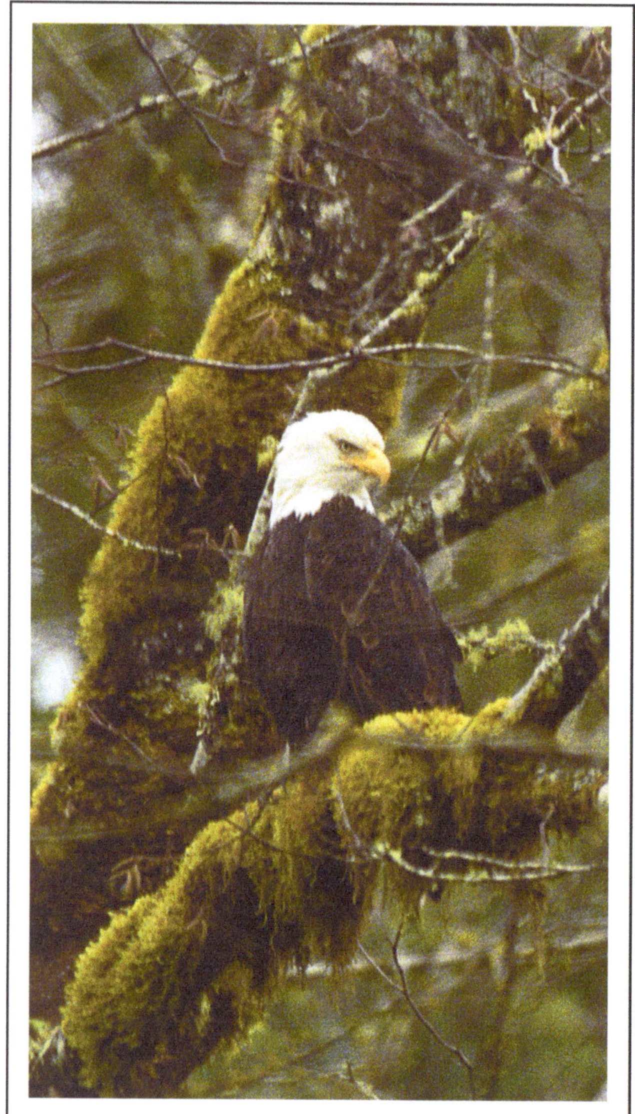

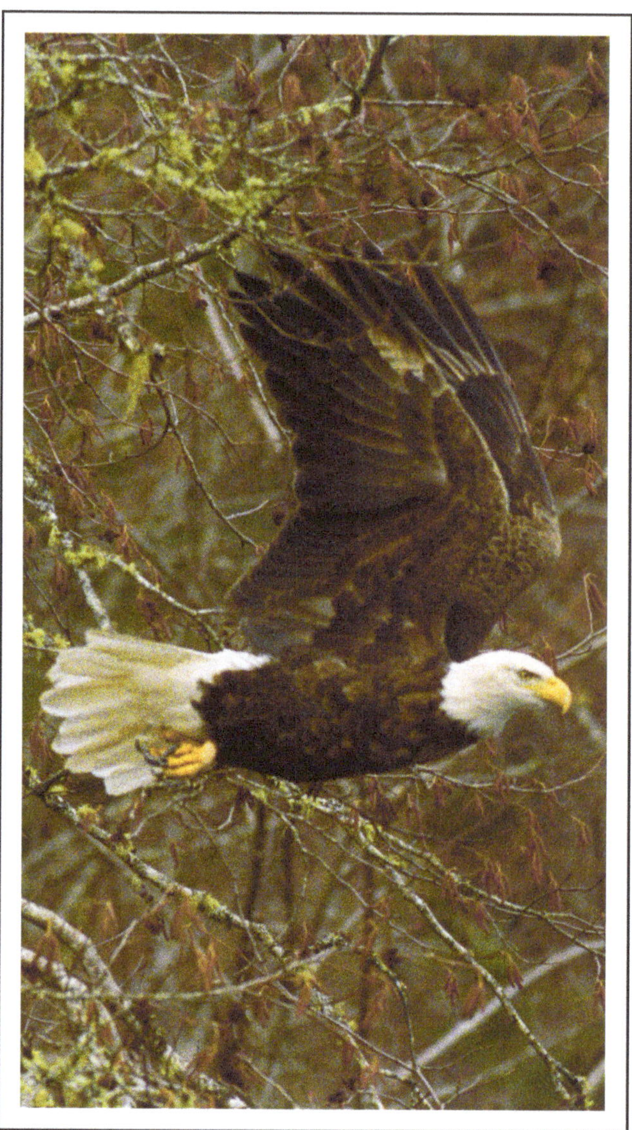

LEFT: *Bald Eagle Lift-Off* (image one) Skagit River in Washington
RIGHT: *Bald Eagle Lift-Off* (image two) Skagit River in Washington

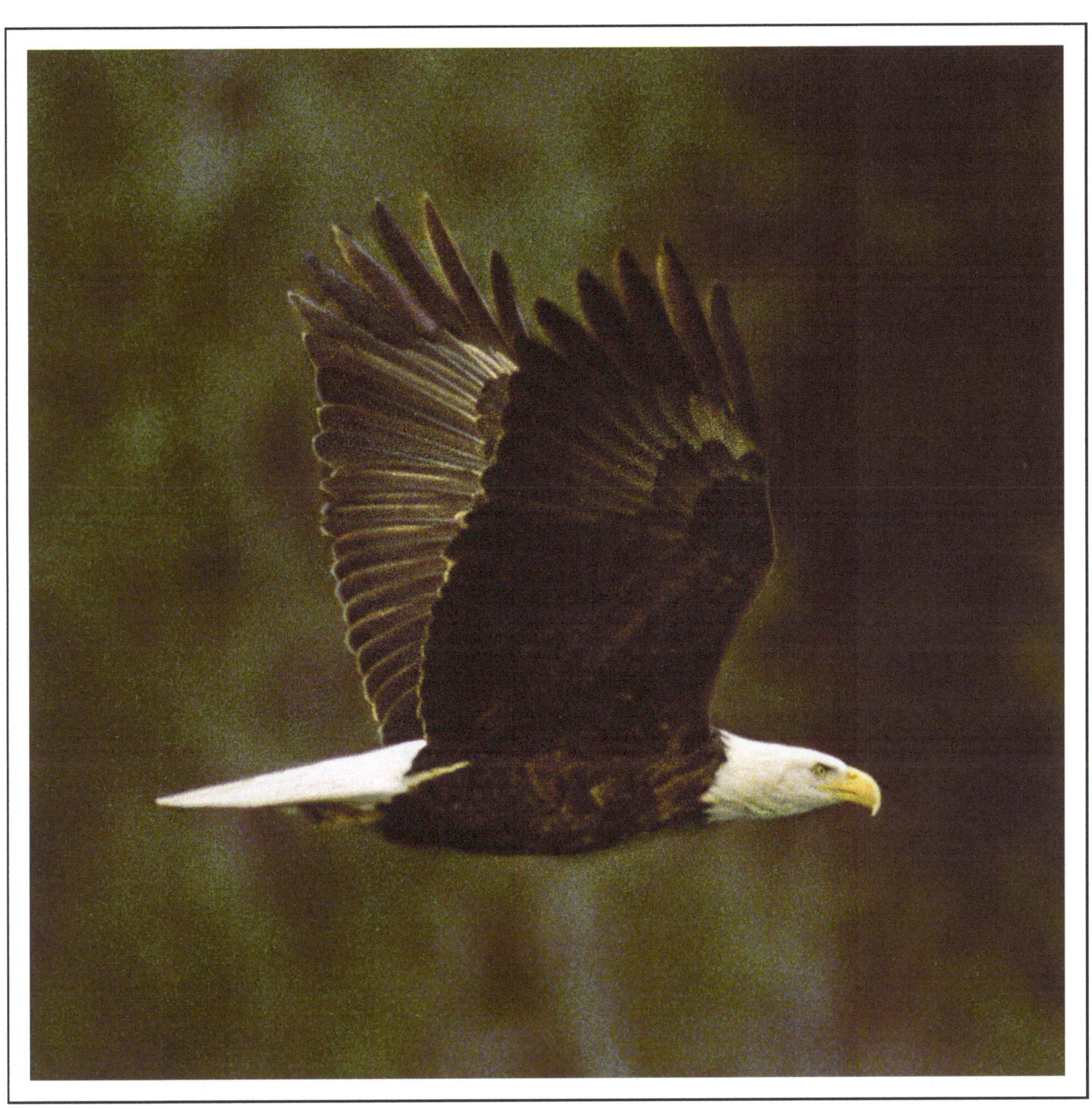

Bald Eagle Lift-Off "I was able to get shots in rapid order of all the stages of flight, from takeoff to a full frame of it soaring wing tip to wing tip." (page 36) Skagit River, Washington

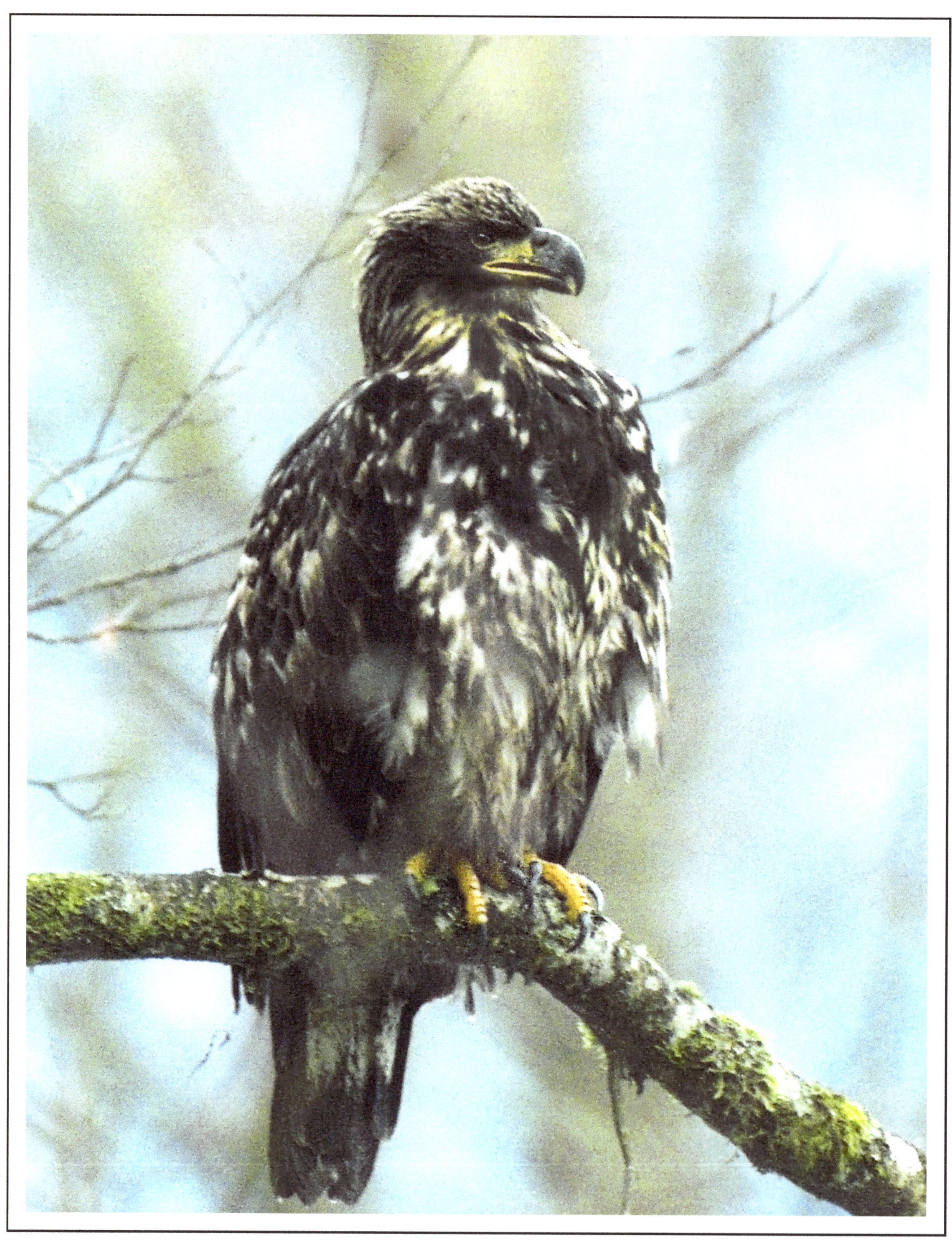

Young Bald Eagle Skagit River, Washington

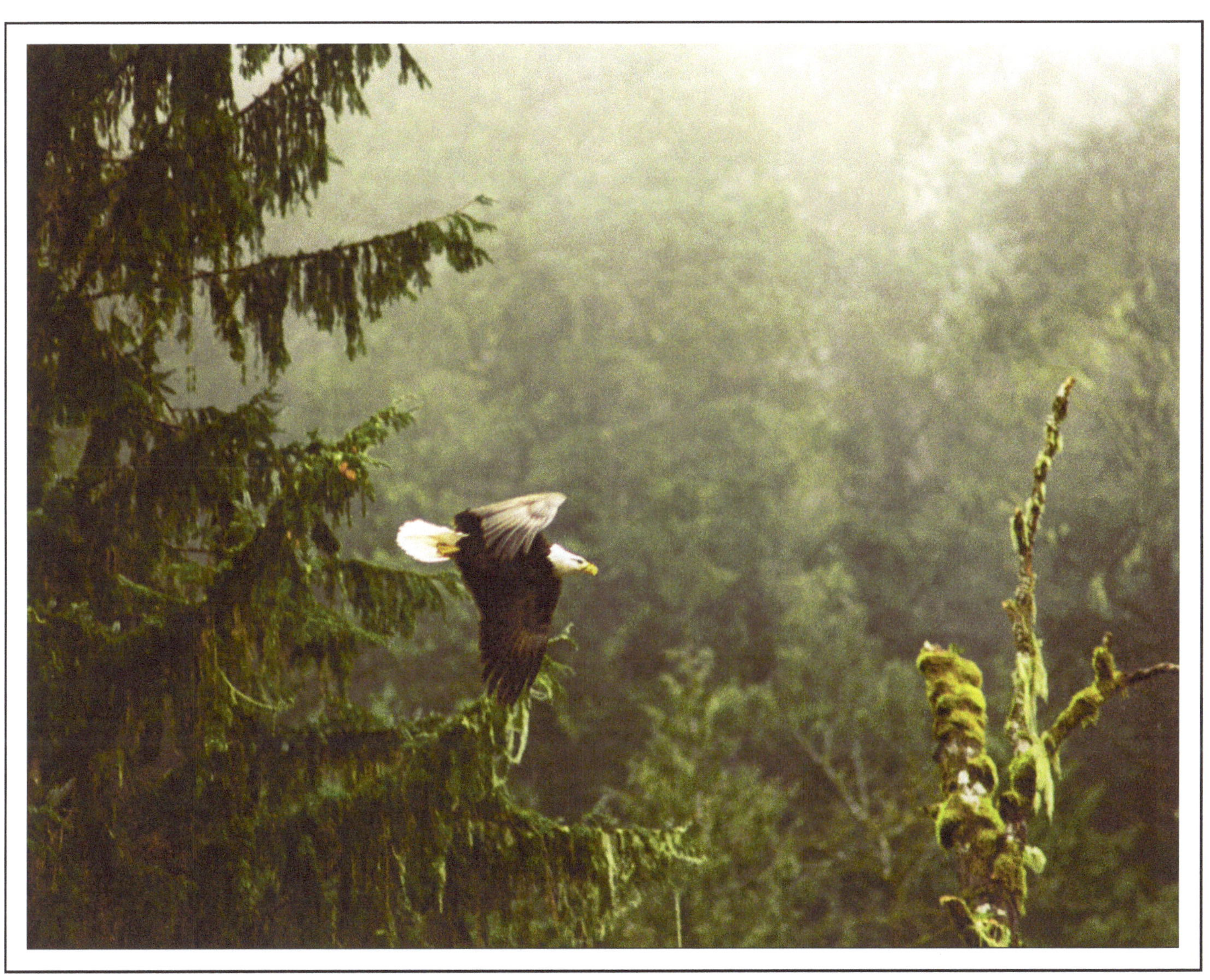

Soaring Eagle "With each eagle that flew through I was overcome with a new wave of emotion..." (page 37) Squamish, British Columbia

CHAPTER 4

A Mountain Lion Story

I arrived in Nelson, British Columbia and checked into the hotel where I was to meet Hugh Robinson, a mountain lion biologist. We planned to go out the next day to track mountain lions in an area that skirted the Cascade Mountain Range near the border between the United States and Canada.

I had met previously with the director of eXtreme Nature, a company trying to fund biological research, and she asked me to do a photo story about mountain lions and what was involved in tracking and collaring them. Hugh had been commissioned to do an environmental-impact report, and he agreed to let me tag along. This was a great opportunity for me to see mountain lions in the wild.

I was a bit anxious and worried about being able to keep up with the biologists and trackers in the wilds of Canada and the Cascades. The process of treeing a mountain lion was basically this: If we found new tracks in the snow, the dogs would be released and typically would chase the mountain lion up a tree. At that point, the dogs would usually start barking loudly and rapidly, signaling us to run quickly to wherever they were. I had heard everyone runs very fast through snow to get to the dogs and the treed lion, and frankly I was afraid I wouldn't be able to maintain the pace and might get lost. I slept fitfully in my hotel room the night before I was to meet Hugh. A scolding voice in my mind kept repeating, *You don't want to get lost*, and this voice kept waking me up.

Hugh arrived early. His calm, modest, but confident, demeanor was reassuring, because at that point I was a nervous wreck. I hoped my anxiety wasn't so apparent that Hugh would pick up on it and begin to question his decision to take me along. He informed me that we would be riding on snowmobiles along the edge of a mountain lion habitat in a valley, looking for new mountain lion tracks. Two other trackers called hounds-men had been out in the wilderness since sunrise scouring the hillsides for any evidence of new tracks. Stu, one of the hounds-men, had showed Hugh what he thought were new tracks, and Hugh used his radio tracking equipment to verify that it was not a mountain lion that had been previously collared.

I began to feel excited that this was actually happening on my first day out. The thought of seeing a mountain lion set my heart racing. I had always been afraid of mountain lions. Most trailheads in California have a sign at the start of the trail cautioning you about them. These signs warn, usually in bold letters, "BEWARE OF MOUNTAIN LIONS" and/or "MOUNTAIN LION HABITAT." They also include instructions on what to do if you run into a mountain lion—usually advice such as make yourself look bigger, pick up your children, yell and wave sticks at the animal. I have taken solace in the knowledge that the big cats were nocturnal and the chance of seeing one was about the same as seeing a tornado. Except here I was with a biologist, who would release hunting dogs to abruptly wake the sleeping lion, and drive it up a tree. The mountain lion would not be happy.

When we arrived where the tracks began, Stu let the dogs go. After about an hour, two of the dogs returned to the base of the mountain where we were, but Levi, the youngest dog, was not with them. Apparently Levi was not yet able to discriminate between animals and was eager to chase them all. Realizing this, Hugh and Stu took the snowmobiles and scoured the hillsides looking for him while I was left as the designated dog-sitter. Before the trip, I had purchased expedition-style clothing and boots to stay warm in the below-zero temperatures involved in winter tracking. Now, over the course of two hours, while I waited for the men to return with Levi, I realized that I had underestimated how cold it could get in the Cascades in winter. It was the kind of cold that permeates your body and leaves you wishing for anything to happen to distract you from it. Even an alien encounter where I was whisked into the "mother ship" would have been preferable to this cold. Luckily, they found Levi where the road bisected a trail. We were done for the day.

On the way back to town Hugh and I compared wildlife stories. I had been a wildlife photographer for about seven years by then, so I was not a novice about the adrenaline rush that results from large animal wildlife encounters. Hugh told a story in which he startled a cougar (a word used interchangeably with mountain lion) who had been sleeping. Fortunately, the large cat ran off without incident. I related the story of how a large grizzly bear had run up a trail I was on, headed straight towards me, then veered sharply and ran past me. (This is the story I tell in chapter 2, "Grizzly and Brown Bears.")

We then described encounters that made us both laugh. The encounter I described didn't involve predators; rather, it was about the moose and the moose hideout, (one of the stories in chapter 1). Hugh's encounter involved startling a female moose with her calf, which resulted in the moose chasing him through a meadow. Just when he thought he had escaped, she stuck her head through the bushes and chased him some more! Finally, he got away safely but it was a close call. We laughed until our faces turned red. It reminded me of the adventure movies where two explorers show each other their scars from wild animals, but in our case it was our bruised egos we were sharing about.

The next time we went out tracking, two days later, was uneventful. We scoured the hillsides looking for tracks. As my tracking ability improved, I was able to distinguish different wildlife tracks. We also found older cougar tracks where a female and her two cubs appeared to have played in the snow, and then decided to hide in an abandoned, dilapidated cabin.

The fourth day started out relatively calmly. A dusting of snow from the previous night was on the ground. Hugh and I had been on the snowmobile in an area we had visited before, and he asked me if I would drive the snowmobile while he walked down the mountain looking for cougars. I felt confident using the snowmobile to drive down the mountain, because I had ridden snowmobiles in New Jersey. But as I rounded a corner, I cut the turn too sharply, and the snowmobile fell over. Now I was on the side of this mountain edge with the snowmobile turned on its side and no one around to help.

I knew that three collared mountain lions had been in the area, because Hugh had detected them earlier with the radio receiver. This also made me anxious. After struggling to budge the snowmobile without success, I discovered I could use gravity to push it down part of the hill. I then took a large stick and jammed it under the snowmobile. Using all the strength I could muster, I was able to leverage the snowmobile upright.

My adventure was not over though. I took a wrong turn into a dead-end fire road, and I didn't have enough space to turn around. I sharply turned the steering to its maximum, then standing on the front of the skis and reaching over the handlebars I gently used the accelerator mechanism on the handlebars to inch the snowmobile forward. But while doing so, I partially edged the front of the snowmobile over the lip of a cliff. As I straddled the front of the snowmobile and looked down at the bottom of the mountain below, I thought, *You're an idiot, Pat.* Carefully, I slid over the top of the snowmobile and back onto the firmer ground of the cliff edge. I couldn't help but laugh at my predicament—I could imagine the headlines: "Woman photographer kills herself by backing herself off a cliff." Finally, mustering up superhuman strength, I lifted the snowmobile in the opposite direction so it was firmly on the road and no longer hanging over the edge.

Still my trials and tribulations were not quite over, because I flooded the engine and had to wait another fifteen minutes before restarting it. I then headed out to pick up Hugh. He had been waiting at the bottom of the hill for quite some time, and was a bit anxious about what had happened to me. When I told him my story about the snowmobile, I imagined he was a little less confident in my abilities. My macho, wildlife photographer credibility was now tainted, and I was close to being placed in the "wimp" category.

On his way down the mountain, Hugh had found fresh mountain lion tracks and they were going to release the hounds to follow the animal's scent through the wilderness. Within ten minutes of releasing them, we heard aggressive barking up on the mountain that turned into even louder rattling barking. Hugh and Stu undressed down to their shirts, pants, and boots, and then Hugh with his large pack led the way up the steep snow-covered hillside after the dogs.

This was it: this was where all my intense preparation just about managed to kill me. I had on my two pairs of socks, steel shank boots, two pairs of stretch pants, sweatpants, plastic outer pants, three shirts, two sweaters, Gore-Tex jacket, sweater dickey, and gloves, as well as my large backpack containing my camera, tripod, two lenses and survival kit. I made it about forty feet and could no longer lift my legs. The area where the dogs had run to was a steep uphill climb, and the snow was about waist deep the entire way. I was sweating profusely, panting, and struggling to continue up the hill, and I was finally reduced to crawling on all fours up the mountain. Hugh saw my predicament and grabbed my pack. Now he was carrying his own more-than-fifty-pound pack and my forty-pound pack, and he still managed to make great time up the mountain. It was a lot easier for me without the pack, but I had to alternate between walking and crawling my way up the mountain.

Then it happened. There, standing high up on a branch in a tree was a beautiful, large, approximately three hundred-pound mountain lion. Her size and the magnificent details of her head and body left me speechless. The dogs were in a pack barking at the bottom of the tree. I grabbed my camera equipment and started shooting. The mountain lioness was on the side of a mountain in a very high tree. It turned out she was from the United States—a "Yank"—but Hugh had no way of knowing she had been collared. Later, Hugh told me he wouldn't have darted the cat even if she hadn't been collared because she was too high up on the side of a mountain and doing so might have endangered her. (Darts are syringe-like containers filled with tranquilizing medicine that's shot into an animal via a tranquillizer gun, which temporarily sedates the animal.)

All along I had been taking shots of Hugh doing different activities involved in tracking mountain lions. I had taken photos of him using his radar equipment and riding on his snowmobile. I had also taken shots of Stu managing the dogs and close-ups of the radio collars. Now I was about to complete my photo story with actual cougar photographs. This was the female cougar with kittens whose tracks we had sighted two days before. I stayed there photographing her for about an hour. Stu had already taken the dogs and headed down the mountain. Now, Hugh turned to me and said I could stay with the cat as long as I wanted; they would meet me down below. I looked at the mountain lion, which looked angry and agitated, and thought, *It's OK, I'm done here*. I felt like a wimp, but Hugh didn't tease me about my failings as a mountain lion tracker. He said the cougar probably would stay in the tree another hour or two, because of how stressed she was.

This experience was amazing, and the mountain lion's image in the tree remains etched in my mind forever.

I've always thought that mountain lions do a great service for us by keeping the deer population in check. This in turn reduces deer-involved motor vehicle accidents and deer-tick bites which can cause Lyme disease. Now, after learning so much about mountain lions from Hugh and experiencing them firsthand, my fear of them is greatly reduced, and I am truly grateful for their presence in our wilderness areas.

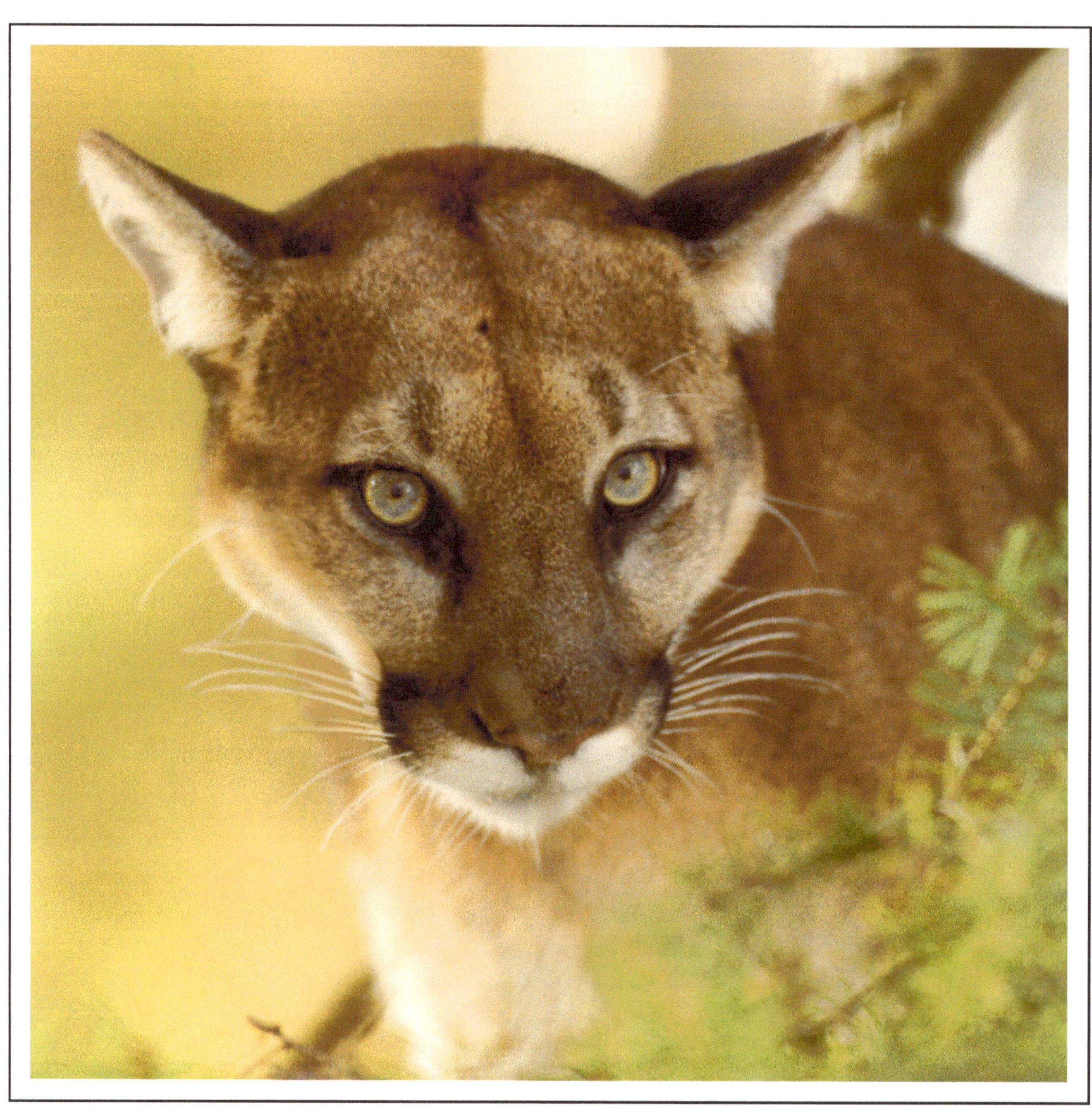

Mountain Lion "There, standing high up on a branch in a tree was a beautiful, large, approximately three hundred-pound mountain lion." (page 47) Cascade Mountain Range

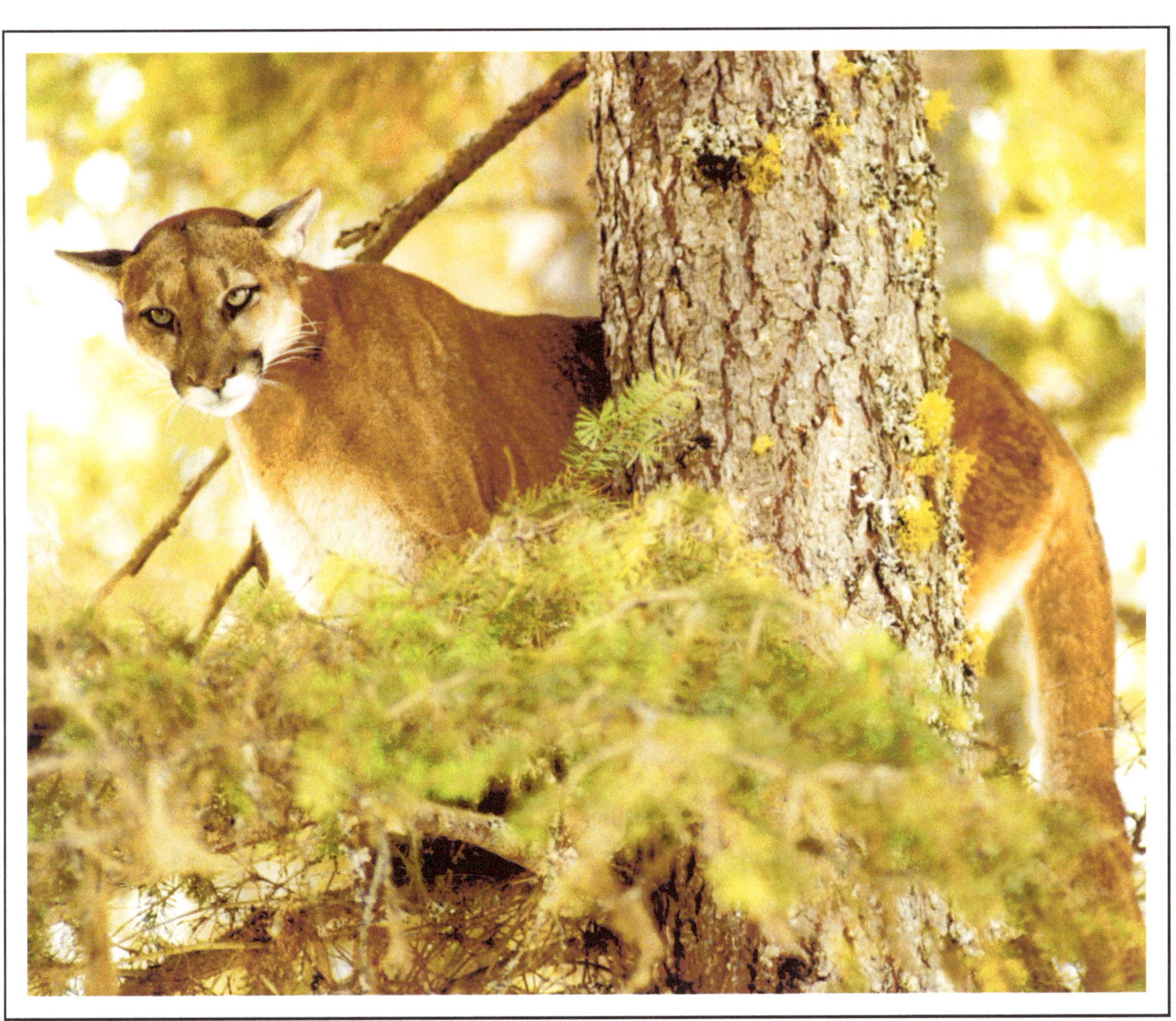

Mountain Lioness Cascade Mountain Range

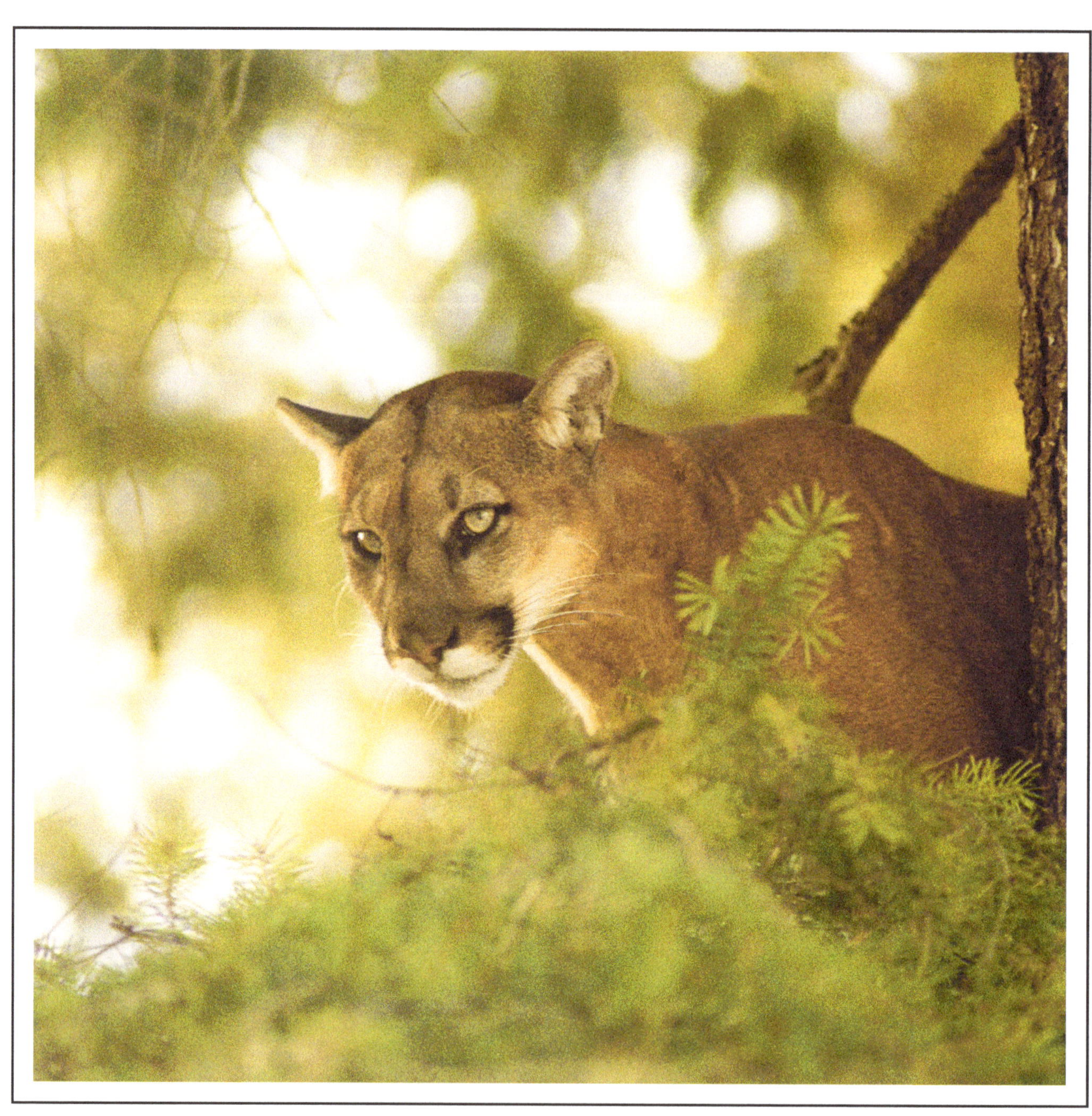

Mountain Lion at Sunset Cascade Mountain Range

CHAPTER 5
Black Bears

Attempting to capture a wild black bear on camera is not an easy task. First, you have to find the bears. Then you have to find them in enough light to make their dark coats visible. Then they need to be far enough away, preferably preoccupied, where they do not take off immediately as soon as they see you—which they usually will. In my experience, black bears typically are not out in the daytime; they like to forage for food after dark. All this makes "black bear paparazzi" work very difficult. I also don't use a flash in my work and I don't carry one with me. Given these bad odds, I have been fortunate in the black bear photo experiences I've had.

One of my first experiences photographing black bears involved a young bear eating berries along the side of the road in Yellowstone/Grand Tetons National Park. It was in the late fall of 1998, when many animals were fattening up for winter. The bear was busy eating berries and resting at intervals. This black bear was small, weighing less than three hundred pounds and seemed to be very young. At intervals, he would lie down in the grass by the side of the bush and bury his head under his paws. Periodically, he would go back to eating berries. Unbothered by my presence, the bear continued his foraging while I photographed him.

Being so close to the road, however, the bear started attracting attention. Our peaceful encounter ended when a dozen or more vehicles stopped and people piled out, causing a "bear jam," a term I would become very familiar with throughout the years. People parked all over, blocking the road and surrounding the bear. I remember one member of this "bear paparazzi" being so aggressive he went as close as four feet from the bear with a throw-away camera and hopped around taking pictures. Surprisingly, the bear tolerated this individual without incident. Fortunately, the park ranger arrived and broke up this potentially dangerous situation, moved the crowd back, and allowed space for the young bear to move on.

On the same trip, while driving through Nevada on an interstate highway, I was tailed by a man who had decided to follow me. He drove a car that was a wreck, all dented up and painted an odd, royal blue. Trying to distinguish my normal female apprehensiveness from an actual threat, I sped up and then slowed down. The person in the dented blue car did the same. Bingo! I wasn't being paranoid, and I now knew that if I didn't lose this persistent threat before settling in for the night, he could become a real problem. "*Serial killer!*" kept flashing in my increasingly alarmed mind.

One of the hardest parts of my job as a woman photographer is driving alone late at night. I feel vulnerable to any deranged human predator who might be roaming the highways and for this reason I try to avoid traveling on the interstates after dark.

As I mentioned earlier, I try to look like a man while driving around. I try to appear tough when people look at me. I behave the same way when I hike out in the parks in the early morning or late evening. I also wear many layers of clothing, don't brush my hair, and make it look like I'm carrying all my clothes on my back. I've intentionally cultivated this "wild and living in the woods" look. The message I deliberately convey is "*stay away.*"

In this instance, the person following me was starting to scare me. He looked deranged. One advantage a woman has when she's driving a car on a major highway is access to the women's bathroom. I decided to avail myself of this advantage, and I quickly developed a plan to extricate myself from the danger I was facing. I stopped at a gas station with a large store and a fast food restaurant attached. Fortunately, it was a crowded station. I parked close to the entrance so I could make it to the bathroom without incident. But sure enough, the stalker pulled up next to me with his car door open. His eyes were wild, and he was yelling loudly, "*Get into the car.*"

I yelled, "*Sure, I'm going to jump in the car with you! You're crazier than I thought.*" Everything seemed to be a blur—I was as afraid as when I encountered my first bear. I then ran into the market and made a beeline to the bathroom.

The time in the bathroom seemed to last forever. Ten minutes went by. I kept asking people coming in about "nutty guy" and if he was still out there. Twenty minutes. Thirty minutes. Finally, after forty minutes he was gone. At last I was able to relax: the predator had lost interest in his prey; 'crazy guy' had moved on. To make sure I didn't run into him again, I ate at the restaurant and read books in the bookstore until I calmed down.

Why I didn't call the police, I don't know. Cellular phones weren't around yet and I guess I was leery of using an outdoor pay phone booth. I feel bad about that to this day, because I really sensed that he was a very dangerous person. His next victim might not have been as lucky as I was. This traumatic experience underscores why I'm so apprehensive about traveling alone after dark, especially on interstates, where the gas stations are often more deserted and if your car breaks down, you could easily become vulnerable to any psychopath who happens to be on the same road.

This aspect of my job has always triggered a great deal of trepidation in me. I've had to adjust many of my trip itineraries out of concern about encountering potentially dangerous people. I've wasted camping fees and good campsites because of suspicious or unusual-looking neighbors. Although it's maddeningly inconvenient, I'm prepared to abandon a site when my instincts and intuition trigger an alert. A person dressed in survival-type army gear wearing camouflage face paint and carrying a large hunting knife when it isn't hunting season should set off the alarm bell!

Motels can sometimes be equally as frightening. I've been at motels where people were hanging out, drinking, and carrying on like it was Mardi gras in New Orleans. This would not be such a problem if they congregated indoors in a bar, but sometimes these parties took place in the parking lot next to my room. On one occasion I requested my money back from the desk clerk. When she handed me my money she actually said, "Dear, I wouldn't stay in this motel either."

Another time at a motel, a much disheveled, inebriated guy blocked the stairway to my room. His arms were stretched from rail to rail and he wouldn't let me pass. I couldn't believe it. When he didn't move after I asked nicely, I channeled my courage into my voice and yelled, "Move!" He was taken aback and got out of my way. With adrenaline surging through my veins I ran up the stairs. Feeling lucky the yell worked, I proceeded to my room. When I turned around, I realized he was following me swaying in his drunken stupor. I flew into my room, slammed the door, locked it and stayed tight all night. Having left my bags in the car, I carried only my camera gear. I had no clothes or toiletries. I spent a restless, funky night sleeping in my clothes. But I wasn't about to go back outside until daylight.

Though I will not be deterred from doing the work I love passionately, I have learned to incorporate a mild and appropriate level of paranoia in my day-to-day functioning when on photo trips.

This brings me to my next black bear encounter, in the Sierra Nevada Mountains. I was in Kings Canyon National Park on one of my many trips alone, looking for wildlife. I was spending a week traveling through the three national parks—Yosemite, Kings Canyon and Sequoia—and camping where the wildlife was. While camping in Kings Canyon, I got up before the sun rose and stopped at a meadow filled with blooming flowers. I parked and hiked out to the meadow. I spotted a fawn and a doe and started following them. The morning light was beautiful and I got really involved with my work.

After an hour the light changed, and I started walking back. I had crossed over a log bridge on my way out to the meadow. On my return, the bridge now had large, wet bear tracks on the logs. The meadow was small, so I knew the bear was close by. It can be scarier knowing a bear is close by, probably observing you, when you have a hike to get back to your vehicle, than it is to actually see it. Do I try to sneak back to my car or use the "'Hey bear!' method"? Again, with all the gusto I could muster, I began yelling and clapping. I even added singing: "If you're happy and you know it clap your hands." Maybe he would think I was a crowd of people. Fortunately, it worked—the encounter never happened, and the bear got an impromptu concert.

In the summer of 2002, I had another bear encounter while hiking in the high country of the Sierra Nevada Mountains in Yosemite National Park. Late in the day, I was hiking with my husband on a remote trail through dense woods. Farther up the trail, we spotted a large black bear coming in our direction. He was lumbering along alone and looked to be over five hundred pounds. That was about all I can remember; my anxiety and body did the rest. Clicking into our "Alaska Bear Trail Mode," we started clapping and yelling "Hey bear!" Luckily, the bear heard us, moved off the trail, and made a wide swath around us—we were able to see him the whole time. Thinking this was unusual for so early in the day, we made a big effort to make more noise hiking the rest of the trail.

A bear encounter is something I've never gotten used to. Especially when I am on a trail and have no car or building to run into, it is always frightening.

Another encounter occurred when I and my friend, Andrea, were in Sequoia National Park in the Sierra Nevada Mountains in the late fall. Our plan was to hike around a small meadow on a designated nature trail. At that time of year not many people were there. When we were ready to start on the trail, my friend opened up her pack and began eating some dried fruit. Hiking with Andrea was like hiking with a traveling food buffet: no shortage of food ever! An elderly couple walked over and mentioned that a bear with a cub had come by earlier while they were eating their lunch, and had stolen their food. They had run away and let the bear have the food.

As they were telling me this, I noticed out of the corner of my eye, what appeared to be the same black bears, the mother and her cub, coming in our direction. I was half distracted by the conversation and knew I shouldn't run away. My friend, on the other hand, the food-holder, ran. I stood my ground and encouraged the couple with me to also not run. Again with all the gusto I could generate, I started yelling "Hey bear!" I began clapping and yelling loudly, fueled by my adrenaline rush. The elderly couple stayed with me, but didn't participate. The bear and cub came within fifty feet of us and then abruptly ran off.

I was so relieved that the bear encounter turned out well. The couple was also grateful and we all sighed with relief. I thought, *I now know what my friend would do in a bear encounter*, as Andrea sheepishly reappeared from behind a redwood tree. I was hoping we had scared the mom and cub enough that they would not return to their old bandit ways because when black bears become aggressive for food in the national parks, they are usually tagged, given a few chances and then destroyed if they don't stop their aggression.

Another experience that resulted in a large number of my bear images occurred one fall when I was in the Sierra Nevada foothills. As I was driving around, I noticed branches in a tree shaking violently. Thinking it was an earthquake, I pulled over. On further inspection I saw it wasn't a quake but a large bear butt, with a small head peeking around from the side that was shaking the branch. The bear had climbed up to the branch to shake it, then climbed down from the tree to eat the acorns that fell. This was luck! I hunkered down quite a distance from the bear and photographed him for about an hour. (This happened before I began using video cameras in my work.)

He was really large—well over six hundred pounds—and was so involved with his activities, he barely noticed me in the distance. The branch he was on was huge and when he would shake, it would almost hit the ground. It was funny and very clever. His rotund weight was a big sign of his intelligence and ability to find food. I remember thinking how lucky I was to see this and to also be able to catch it on film. The bear would occasionally look in my direction and then continue his feast. This image will live on in my mind.

When more people stopped to watch him, the bear moved off. Something I've always noticed is that people really appreciate seeing these animals. No matter who they are, their excitement always pours out, along with their bodies, from their cars, running in the direction of the bear. Just like some celebrities, the bears also run—in the other direction. It's all a healthy balance, and keeps bears and humans apart.

My last and most important black bear encounter occurred in 2007 and involved a mom and two cubs. One early morning I was driving in the Sierra Nevada foothills when I noticed a female black bear and her family by an oak grove. They were preoccupied eating nuts. I pulled over, slowly set up my equipment and started photographing. I was alone and quite a distance from them.

While the mom was intently eating acorns, the two cubs were busy playing and climbing the nearby trees. They would chase each other back and forth through the grass. The darker cub then sat in the grass making funny faces. It seemed as though he had a piece of dry grass in his cheek and was trying to get it out of his mouth. The other cub scrambled around in the leaves, looking as if he was playing in the piles of fallen leaves. I was lucky to stay with them for over forty minutes.

Then one of the cubs, after depleting her limited attention span, thought I looked interesting and wanted a closer look. Mom was not happy, and neither was I. I made scowling faces and clapped and yelled, "No!" (Did I think I was talking to my young daughter?) My anxiety level was climbing; the cub kept coming at me. By this time the bear mom was growling, barking and tapping the ground. We were both stressed, and fortunately the cub abruptly turned and went back to Mom. The bear continued to bark at the cub for a few minutes, and then she and the cub had a moment where they were in a bear hug. I thought, a "bear time-out". It was also time for me to leave.

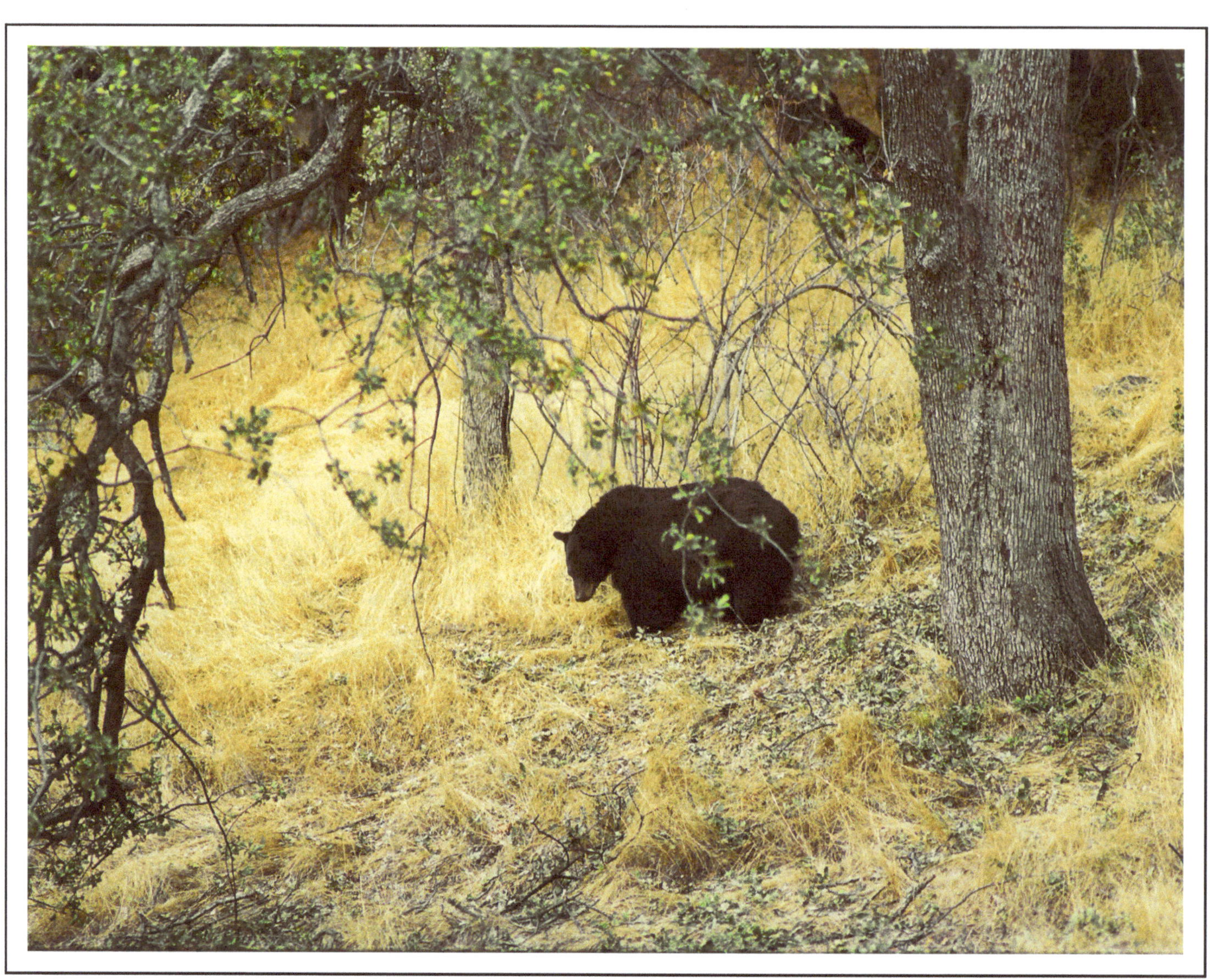

California Black Bear Sierra Nevada foothills

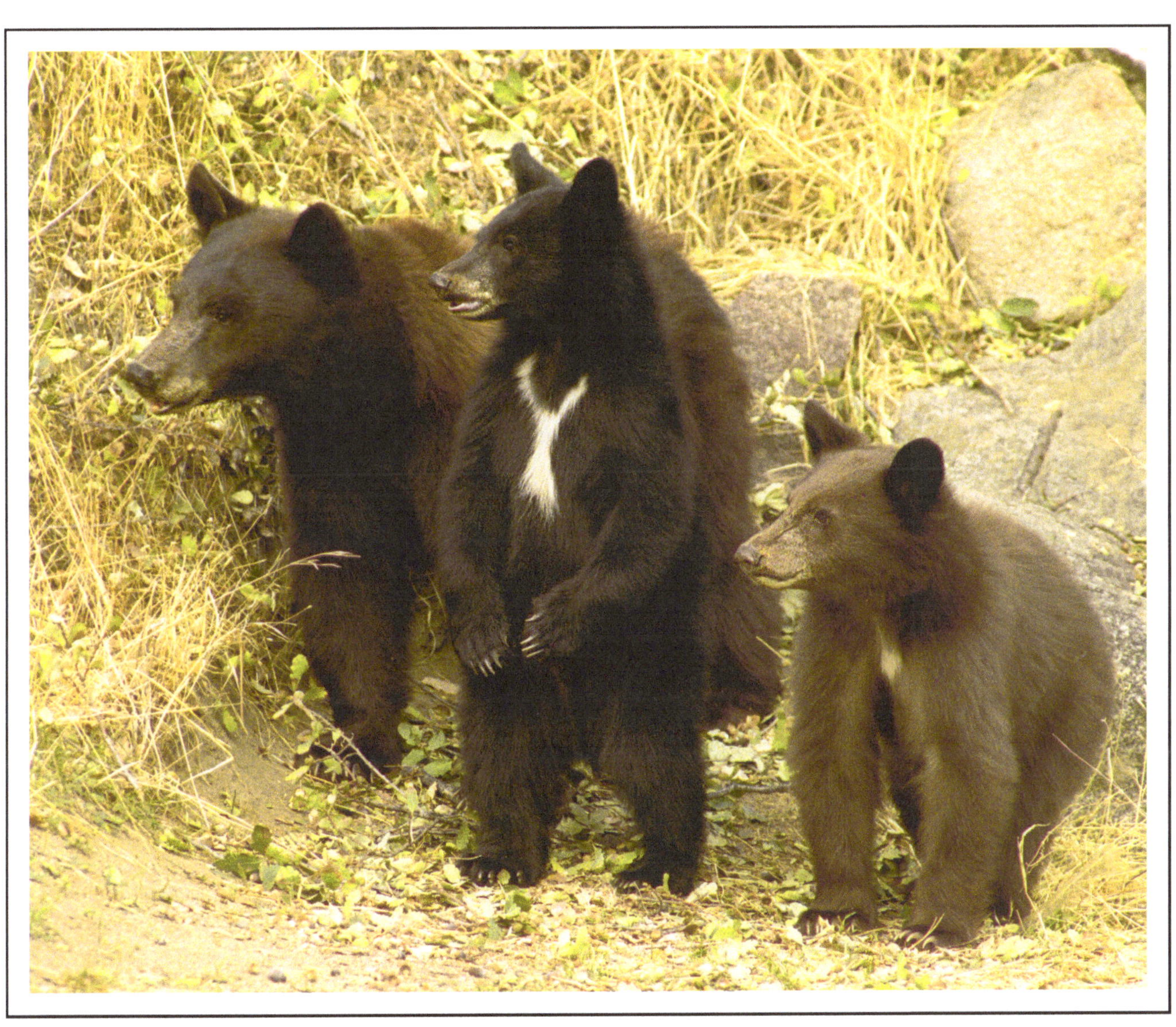

Black Bear Family "One early morning I was driving in the Sierra Nevada foothills when I noticed a female black bear and her family by an oak grove." (page 55)

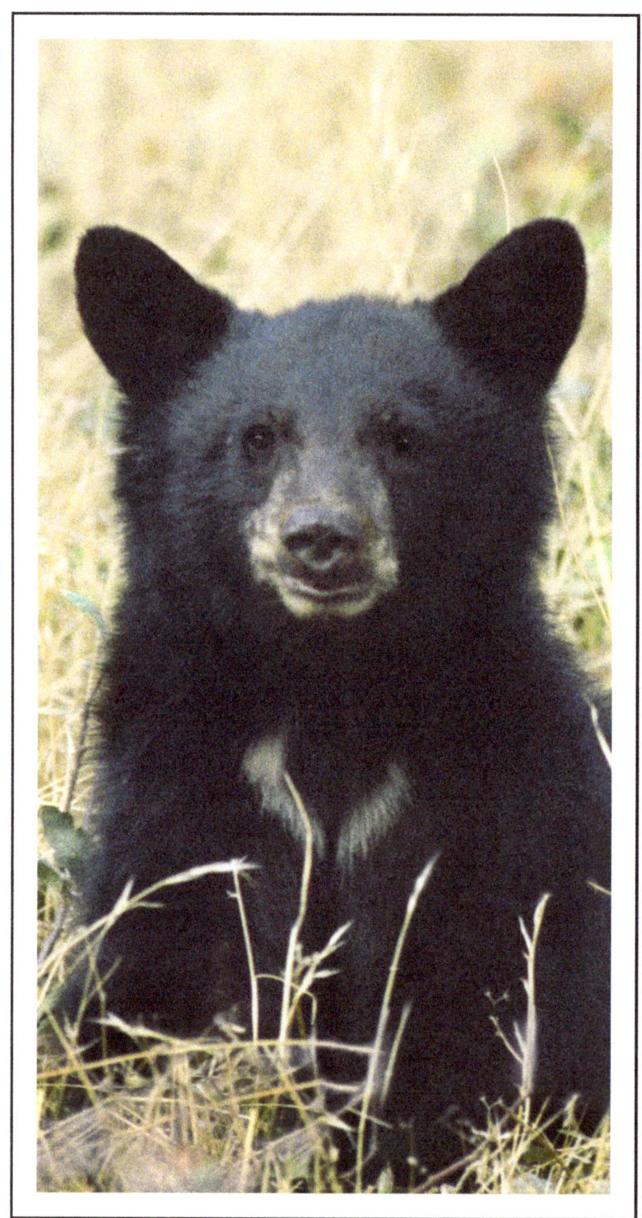

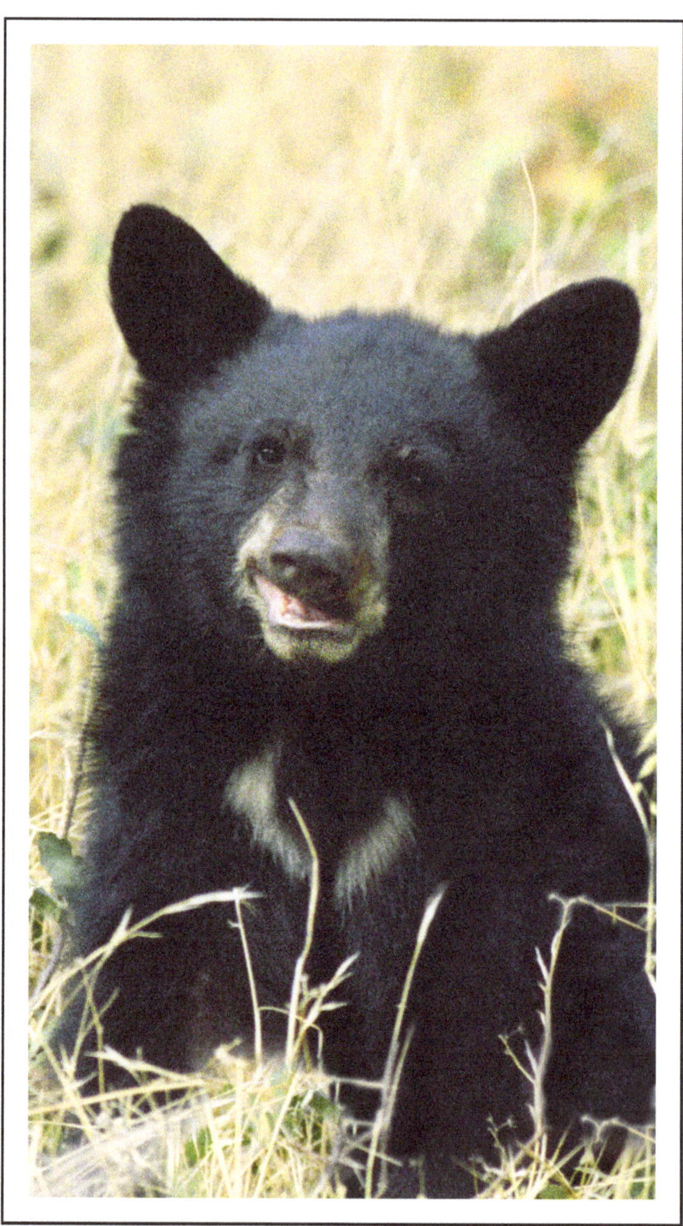

LEFT: *Crack a Smile* (image one) Sierra Nevada foothills
RIGHT: *Crack a Smile* (image two) Sierra Nevada foothills

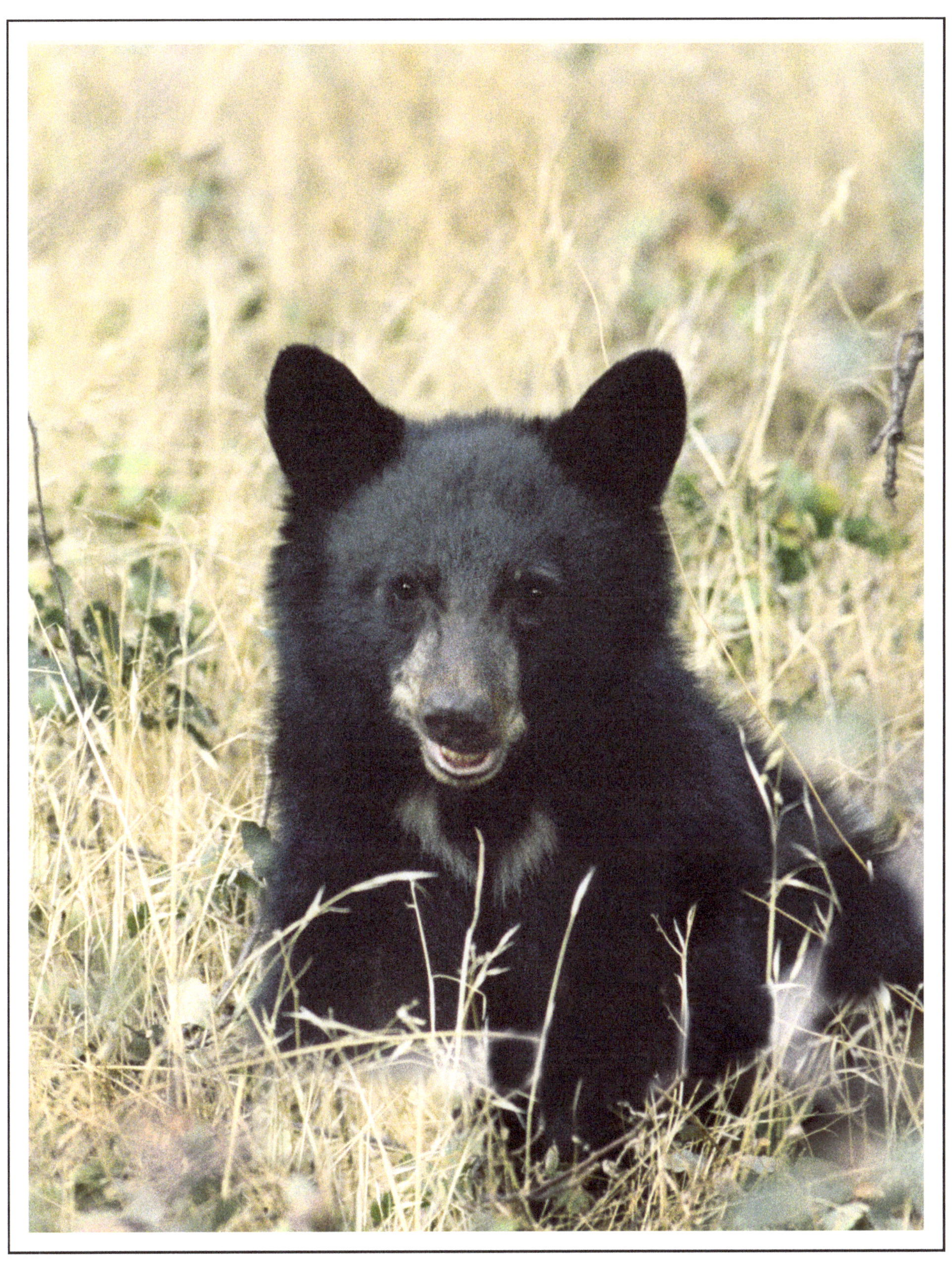

Crack a Smile (image three) " The darker cub then sat in the grass making funny faces. It seemed as though he had a piece of dry grass in his cheek and was trying to get it out of his mouth." (page 55) Sierra Nevada foothills

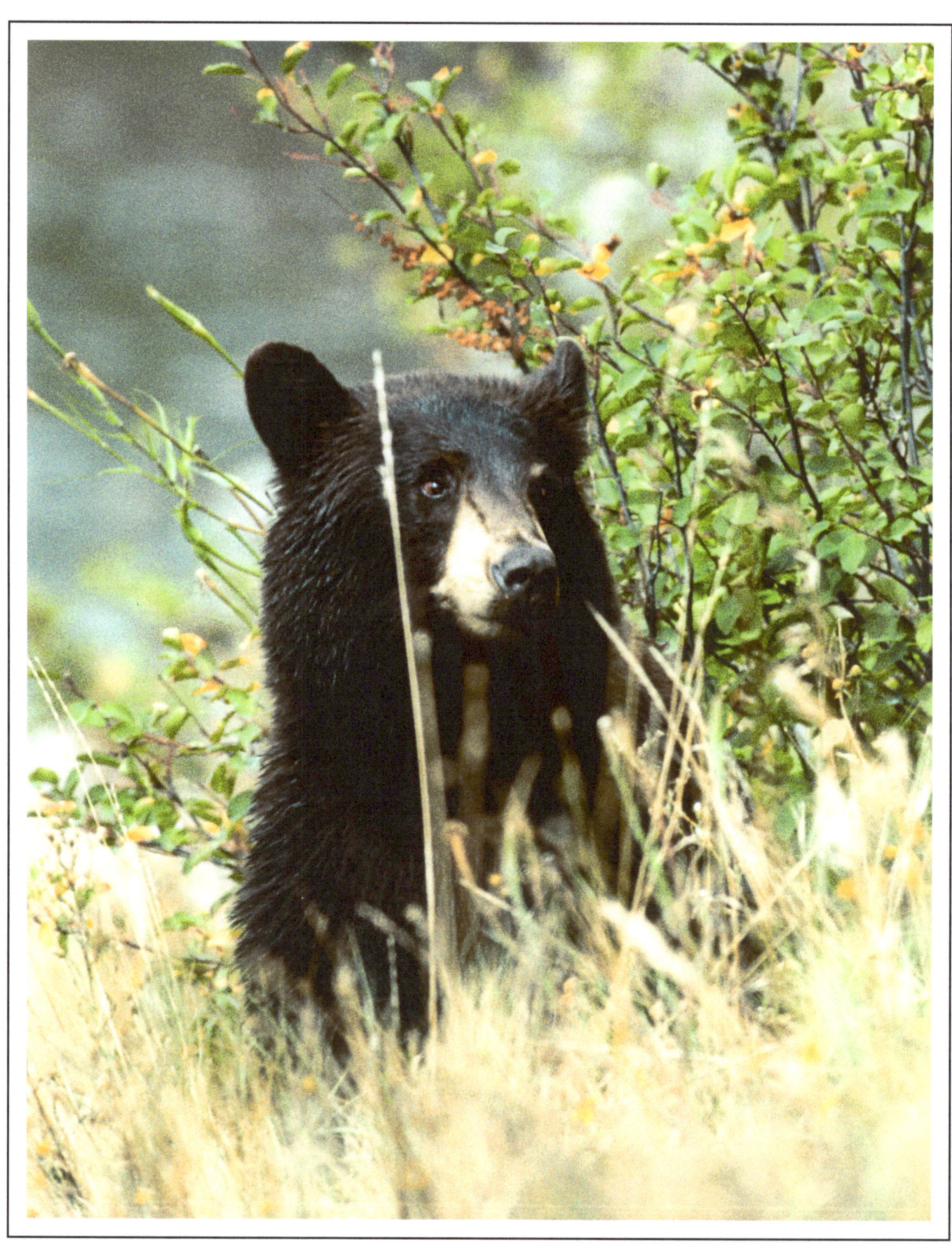

Black Bear Hiding Taken along the side of the road in a national park

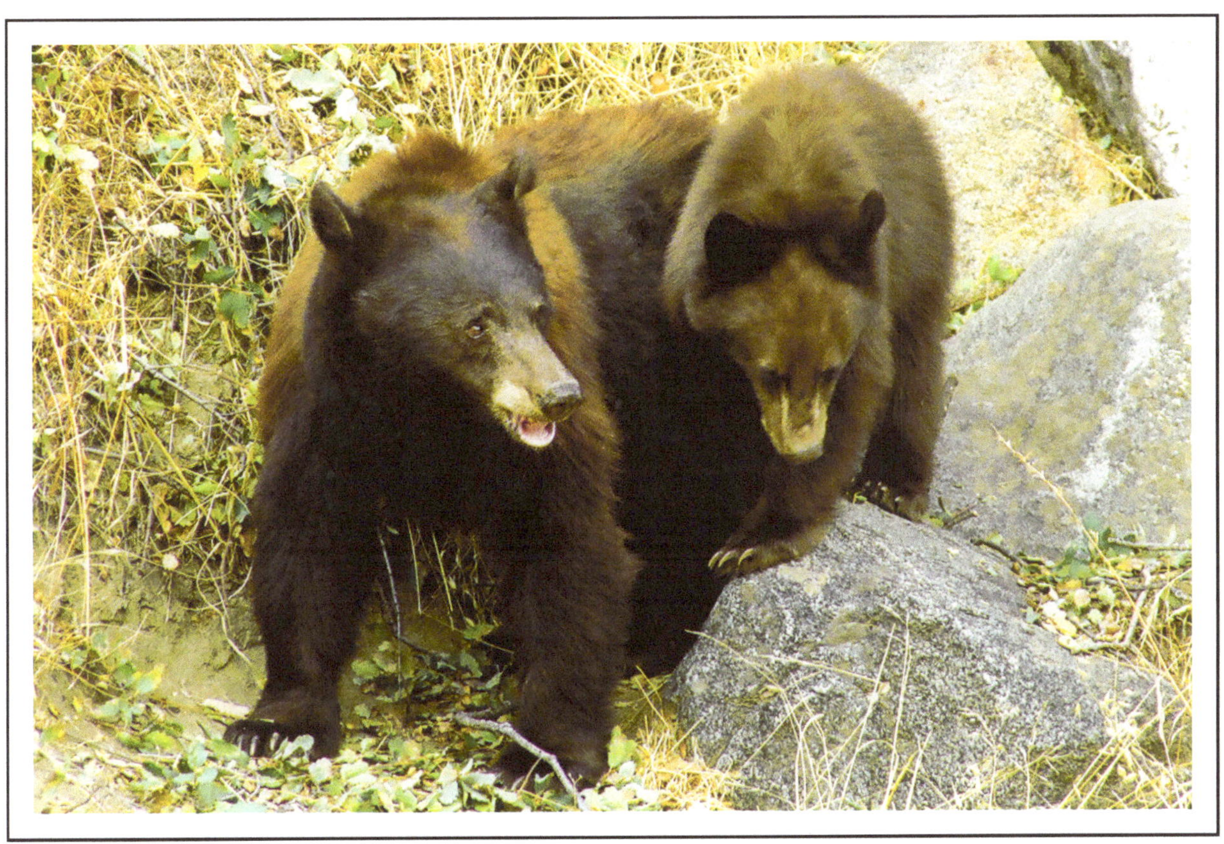

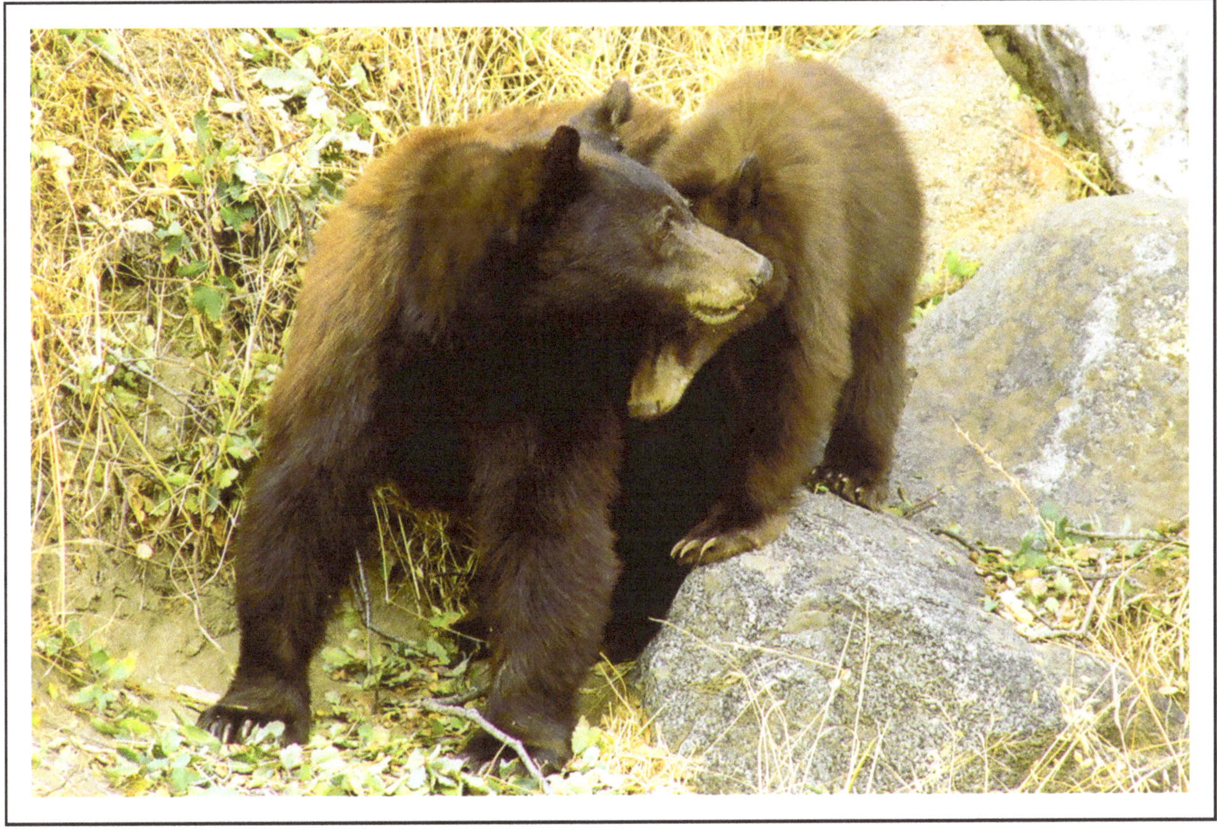

TOP: *Time Out* (image one) “The bear continued to bark at the cub and after a few minutes she and the cub had a moment where they were in a bear hug.” (page 55) Sierra Nevada foothills
BOTTOM: *Time Out* (image two) Sierra Nevada foothills

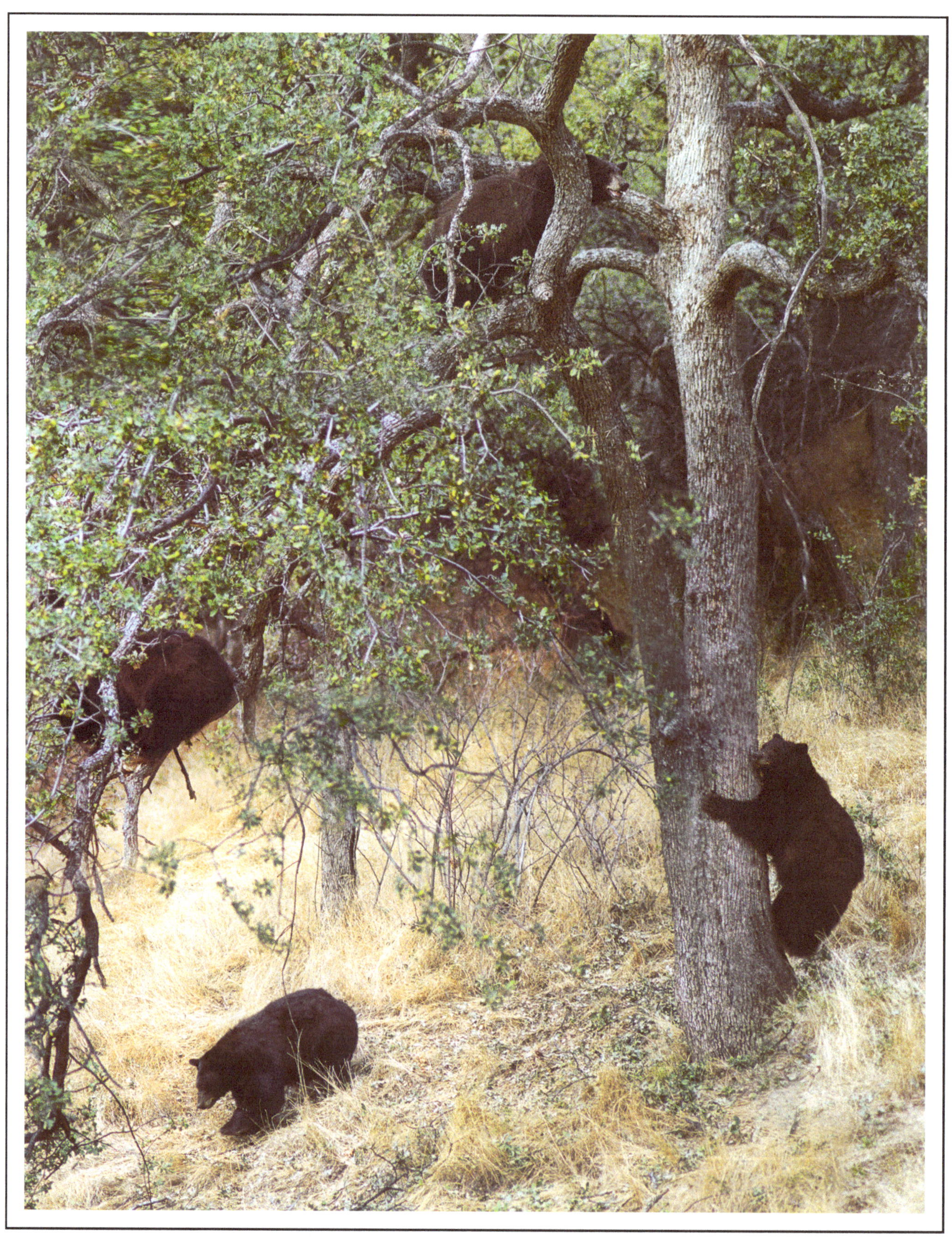

Black Bear Gathering Acorns "On further inspection I saw it wasn't an earthquake but a large bear butt, with a small head peeking around from the side that was shaking the branch. The bear had climbed up to the branch to shake it, then climbed down from the tree to eat the acorns that fell." (page 54) Sierra Nevada foothills

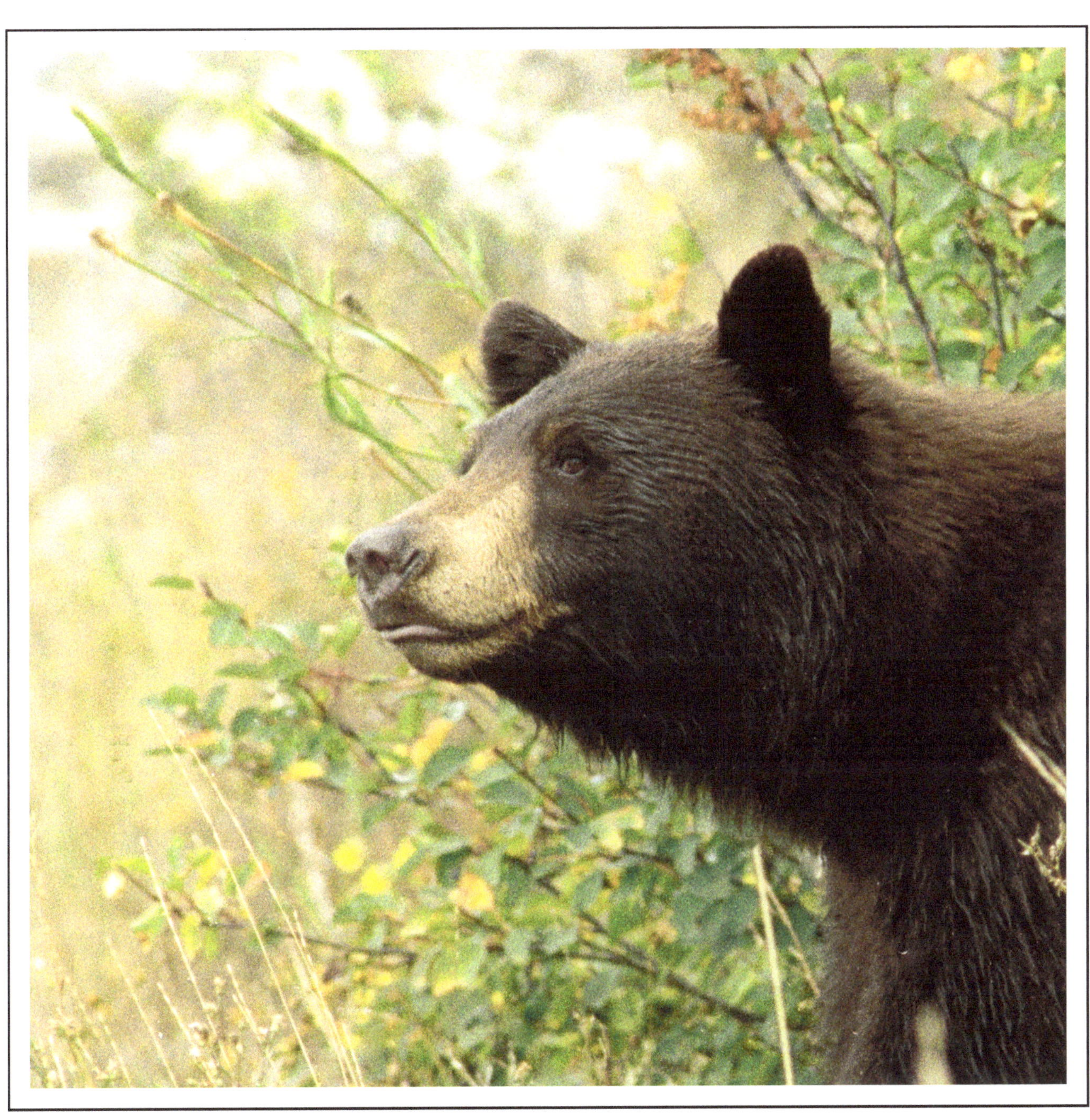

Contented Black Bear "The bear was busy eating berries and resting at intervals." (page 51) Yellowstone/Grand Tetons National Park

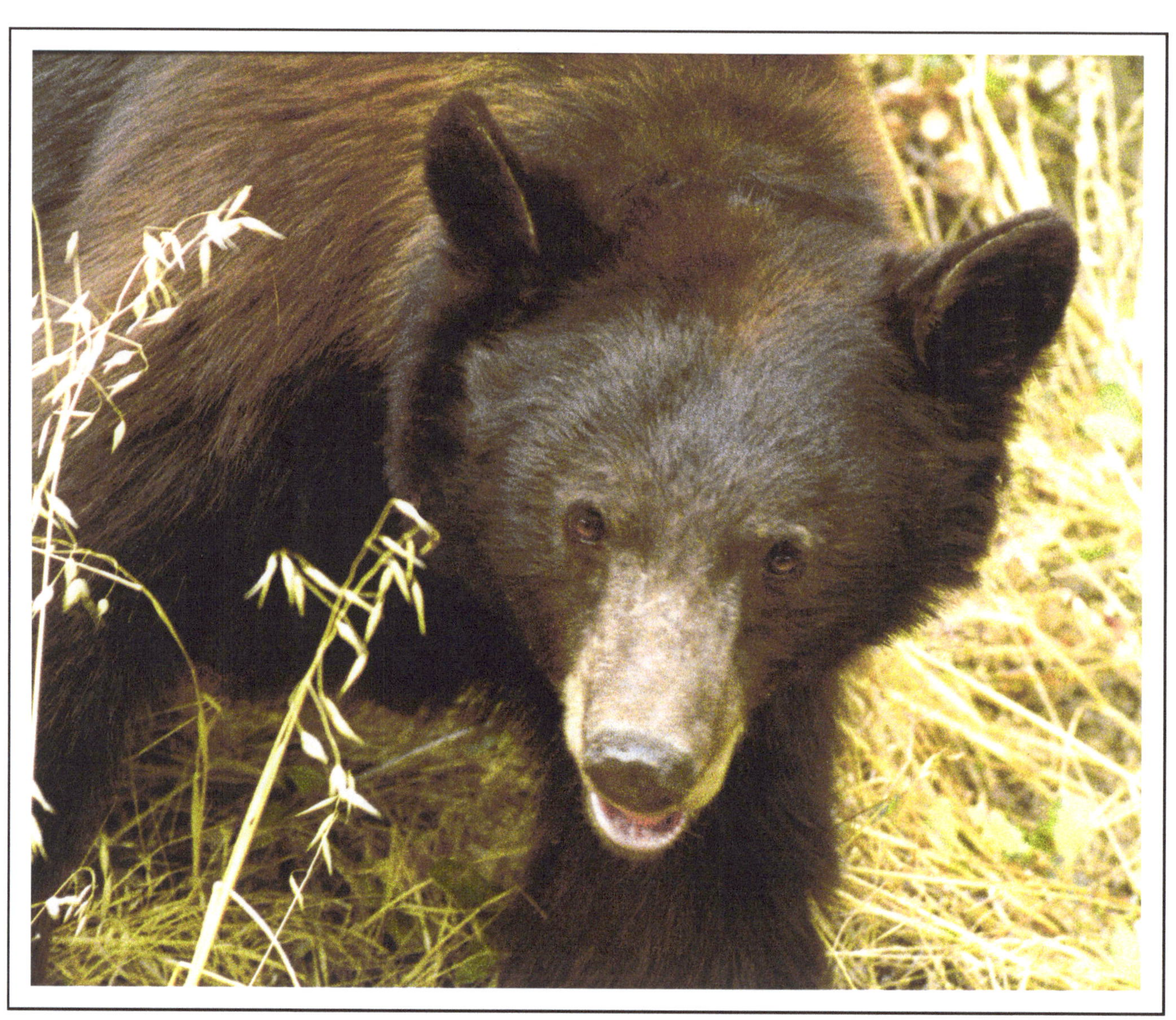

Black Bear Close-up Sierra Nevada foothills

CHAPTER 6
Cry of the Wolves

My pursuit of wolves in Yellowstone National Park has been fraught with many obstacles. Re-introduced to Yellowstone in 1985, wolves have adapted well, proliferated and expanded into remote areas of the park. Wary of humans, they steer clear of busier places and generally keep to their established territories. Also, being elusive animals, wolves generally sleep in the daytime and hunt at night. They blend in with the background landscape, which helps them stalk their prey.

The best time to see wolves is in the winter when their gray fur contrasts with the white snow, but this may mean getting caught in Yellowstone's unpredictable weather. At an elevation of over seven thousand feet, freak snowstorms, even in summertime, and plummeting temperatures of twenty to thirty degrees below zero are not uncommon.

My first trip to Yellowstone searching for wolves was in March 2001. I was driving my van loaded with my boxed fine-art images from Aspen, Colorado, where I had scouted new galleries. I was on Interstate 80 headed towards Yellowstone National Park when it began to snow heavily. The winds blew and the terrain was so flat, that the road would get snow covered, blend in with the hillside, and disappear in front of me. Alarmed and scared, I slowed to a crawl. I kept my window open and, with the freezing air and snow pouring in, I frequently stuck my head out to look for the highway. My pulse quickened more when, in the distance, I saw a car stuck in a field and covered in snow like a statue honoring the road gods. The road had turned, but the vehicle hadn't. Not wanting to be another casualty of the storm, I began hugging the back of a tractor-trailer truck. It was a messy adventure, since the truck's wheels threw slush continuously at my windshield. My wipers partially cleared my view, but they ground in the icy mush, leaving indelible grooves in my windshield. At least I felt safe. Eventually, other cars sucked up behind me, and we formed a caravan barreling along behind the fearless truck driver through the storm.

I finally arrived in Jackson, Wyoming, in the early afternoon and inquired at the visitor center about wolves. They said the best spot to view them was in the Lamar Valley region of Yellowstone National Park. After I spent a day snowmobiling around the southern end of the park, I started my drive north through Idaho to Gardiner, Montana, at Yellowstone's' north entrance. The drive was long and circuitous, through another snow storm, and took all day.

I stayed at a hotel in Gardiner and was up before the crack of dawn for my drive out to the Lamar Valley. As I climbed the winding, steep road in the dark to Mammoth, it was snowing lightly. I stopped frequently for the scattered elk that appeared suddenly in the middle-of-the-road. It took me at least an hour to get to the park.

When I arrived in Lamar Valley, I was shocked at what greeted me. Even though it was still dark, the two pullouts in the vicinity closest to the wolves were full of cars and people with scopes and large lenses. I had stopped next to a pullout, my van's back end partially in the road, when a guy in a ski mask came up to me and told me to move. He insisted that park rangers would soon tell me to do so. I pleaded my case but he was insistent that the rangers would make me move. I could see a group of wolves in the distance, but after further discussion I reluctantly drove off. I was upset, but I thought, *I have two more days here; I'll come out earlier tomorrow and get a spot.*

I spent the rest of my morning photographing other animals foraging in the snow. I kept checking back at the pullouts and returned in midafternoon. The wolves had since moved over the ridge, barely visible, and as a result some people had left. I spoke to a few people milling about. They seemed dazed and excited talking about all the wolf-pack behaviors they had witnessed. Their fervor helped explain the full-to-bursting pull-outs I had seen that morning. During this trip I learned there are many dedicated wolf viewers who closely follow the different wolf packs. Some of these viewers have even quit their jobs and moved to the Yellowstone area to become volunteers for the wolf reintroduction program. They had tracking equipment and were active on walkie-talkies, conveying updates to their colleagues as they monitored the different pack movements and behaviors.

While I hung out with the *wolfies* (a nickname I gave the dedicated wolf viewers), a lone wolf stuck his head out over the horizon, and then appeared on the cusp of the hill, thus giving me a brief glimpse of a wolf. Although the wolf was too far off to photograph, he excited me, and I looked forward to the next day.

The next morning I awoke up at 2:30 a.m. to a heavy snowstorm outside. I remembered the treacherous roadways and decided to wait until the snow-plows cleared them. The snowfall slowed by 9:00 a.m. as I headed towards the park. By then the roads had been plowed and salted and I drove easily to Lamar Valley. Although the pullouts were full, a ranger let me park on the roadside if I agreed to stay with my vehicle.

The Druid pack, the original wolf pack reintroduced in Yellowstone, was on a hillside in the far distance. They were too far away for a decent photograph, but it was exciting to watch them. At least fifteen wolves rested in the snow, with a few pups playing together. The wolves' fur colors were split between dark gray, almost black, and tannish white. It was amazing to see their playfulness and their concern for the pups as they rollicked around. I was starting to get an understanding of people's extreme fascination for these animals. I hung out for a few hours, then, as it started to get dark, headed back towards Gardner.

On the way I spotted another group of wolves, the Rose Creek pack. Three of the wolves walked in the canyon valley below me. I inched my van into a pullout and grabbed my camera but it was soon apparent they were too far off for a good picture. I settled back in my seat and watched, riveted by their behavior. One wolf split off from the other two and made its way up the hillside to the right through trees and around a large boulder, while the other two remained in the valley. A large horned elk, unaware of the wolves, was foraging on the hillside. Suddenly, all the wolves ran at the elk from their positions. Finally alerted, the elk ran and disappeared over the hillside with the wolves in hot pursuit. It was almost like watching a documentary. At this moment I further understood the deep fascination the wolfies had for following the wolf packs. I was giddy as I drove to the hotel, and I hardly slept that night.

The next day was my last in the park; I had to leave midafternoon to drive home. I woke up at 2:30 a.m., dressed, packed, and headed out. Fortunately it wasn't snowing, but it still took me an hour and a half to get to Lamar Valley because the roads were coated with ice. When I got there the pullouts were stuffed with cars, and I tried to inch my way into one. One of the watchers came up to the car and told me the pack had made a kill, and there wasn't any room for my van. With tears in my eyes, I felt like, *Hey I'm one of you, I'm a wolfie—can't I squeeze in?* But, like before, there wasn't any room, and dejected, I moved on.

Driving with watery eyes, I knew I only had hours left in the park and here was my last chance to photograph wolves.

Had I blown it?

I saw the kill site from the road. It had been a large male elk. Coyotes and ravens fed as if in a frenzy on the remainder of the carcass. When I continued over a hill, much to my surprise, on my left was a large gray wolf close to the road. I drove partially off the road, flicked on my hazard lights, grabbed my camera and shot off multiple frames—like a crazy woman. His yellow eyes locked on mine when I paused from taking photographs for an instant. Then he slipped across the road and down into the ravine.

The whole encounter lasted only about a minute, but from it I acquired my two favorite wolf shots. His size appeared much larger due to his thick winter coat, and though it was light out, his yellow eyes seemed to glow. He had a lumbering gait, but seemed spry with no obvious injuries. A few years later, I showed my photos to a ranger who said the tufts of hair around the neck and shoulder suggested a young wolf, maybe a teenager.

As I left for home, I thought, *Maybe not completely fitting in with the wolfies turned out not so bad. With the pullouts full, I was propelled to keep on the move and that gave me this great wolf encounter.* A smile was imprinted on my face all the while I drove through four Western states to California.

I made the next trip to see wolves in Yellowstone in the winter of 2006. To save travel time, I flew into Billings, Montana and rented a four-wheel-drive vehicle, which I drove to Gardner, Montana. The moment I got in my car it started snowing, and it didn't stop for the six days I traveled the park. The snowy weather vacillated between storms, light flurries, heavy flurries, icy hail, and blizzard-like conditions (all of which were not the easiest thing to photograph through).

My first morning in the park was uneventful, without any wolf sightings, but I got some nice buffalo and elk shots. On the second day, I took a snow-cat bus ride down to Old Faithful. Snow-cats are large bus-like vehicles with tank-like wheels that move like a treadmill over the snow. The Mollies wolf pack had killed a buffalo along the route two days earlier, and I hoped to see the kill site. The thought that I might see a wolf visiting it excited me. As it turned out, it was a day of blizzard-like conditions, and the site was covered with snow. Farther down the route, a bald eagle in its winter plumage sat statue-like on a branch. I photographed through the blinding snow and was surprised at the Monet-type effect the snowfall gave the eagle. I enjoyed the snow-cat trip and the eagle encounter, but was anxious to see wolves.

My third day in the park, I came upon a large crowd of wolf viewers on a hillside in Lamar Valley. I parked my vehicle and walked to where the viewers were. In the far distance were between six and eight wolves from the Druid pack. Although the weather alternated between storm-heavy snowflakes and an occasional flurry, I captured some shots of the wolves on the move. The heavy snow gave the images the appearance of an impressionist painting.

I also struck up a conversation with an older gentleman with a white beard who had a quiet, easy-going manner. He looked familiar, and when he gave his first name as Thomas it dawned on me that he was the accomplished wildlife photographer Thomas Mangelson. After his companions confirmed it and I introduced myself we spent part of the afternoon sharing wildlife stories and talking about technical issues with digital and film photography. The wolves were far away and the storm conditions worsened so I called it a day.

On the fourth day I returned to Lamar Valley, where I was told the Druid pack was much more active and had moved to the other side of the valley. I joined the viewers there, and we watched the pack circle an elk. It was exciting, and I was caught up in the enthusiasm of the wolf watchers and the other photographers present. Thomas Mangelson was there, and we settled back into our previous conversation. Suddenly, the pack was on the move—the wolves were trying to bring down the elk. Everyone around me jumped into their cars and began driving to the south end of the valley. The chase was on. I jumped into my car to follow the pack.

At the viewing spot, I had just gotten out of my rental car when all of a sudden my horn started blaring. It blared over and over, even with the ignition turned off. I was mortified. All the photographers, including my new buddy, Thomas, were staring at me with a look of, *Shut that thing off—are you crazy?* Frantically, I tried everything, but nothing worked. Finally, I got in the rental car and drove away, my horn still blasting.

I was so embarrassed. I drove for ten minutes with the horn screaming. I had no idea how to turn it off. All of a sudden, for no apparent reason, it finally stopped. At this point I was overtaken by a fit of giggles, laughing loudly. I started saying stupid things like "Oops! Pat Toth-Smith, professional horn-blower, I mean photographer, horn-blowing wildlife photographer." I laughed the hour's drive to Gardiner. The expressions on the faces of all the other wildlife photographers were etched in my mind, and my embarrassment kept me laughing. All the respect and integrity capital I had earned over the last few days went out the window. I thought, *I'm Pat Toth-Smith, the stupid horn-blowing disturber of the peace.* I had broken the unwritten number one rule of wildlife photography: to make no noise and to not disturb the wildlife. I had not only failed miserably to keep the rule, I had pummeled it.

The worst was yet to come. When I called the car company, the manager informed me that she had no idea why the horn went off and I would have to bring the car back if I wanted assurance it wouldn't happen again. I was looking at a three hour ride: one and half hours each way, at night, during a medium-size snowstorm. Stuck with no other options, I set off for Billings.

The heavy nighttime snow flurries created an eerie landscape. My headlights suddenly illuminated an elk, and I just missed running into her. I over-braked, which threw me forward in my vehicle. Exhausted—I had been up since 2:30 a.m. to photograph the wolves—and with a dangerous roadway before me, I sat up straight and put a double-handed death grip on the steering wheel. I found I stayed more awake when I leaned forward. The storm-bound two-lane highway was dotted with scattered elk and poorly lit. At times my visibility was only twenty-five feet, and I had to slow to twenty or twenty-five miles per hour accordingly.

Frazzled and exhausted, I pulled into the car-rental facility at 10:30pm. After I told the events of my day, the rental car employee, a perky young woman, informed me that an alarm button on the keychain had caused the horn to blast. Worse, it was a safety feature that could easily be turned off by pressing a tiny button on the keychain. Furious that the manager didn't know about the alarm, that it wasn't explained to me when I picked up the car, and that this treacherous drive and awful experience could easily have been averted, I proceeded to the ladies room to regain my composure. I insisted on a new car, to justify my trouble and drive, and instructed the employee to pass on my displeasure to the clueless manager.

After I had drunk a huge amount of coffee, I returned to my elk-filled, ice-covered, stormy roadway and inched my way back to Gardiner. As it turned out this day had been my last opportunity to see wolves; they were absent the remainder of my trip.

In May 2012, I squeezed in a trip to Yellowstone National Park between several art shows. I flew to Bozeman, Montana, rented a car, and planned to set out for Yellowstone early the next morning. Before leaving for Yellowstone, however, I came down with a severe flu. The flight was miserable. I alternated between severe congestion and chills from a fever. Knowing I would be in the park for only three and a half days, I forced myself up at 4 a.m. the first day and traveled from Bozeman to Yellowstone National Park, arriving at Lamar Valley at 7 a.m. I was saddened to learn from the wolf watchers that the Druid pack was no more. I suffered at that thought. Memories of the pack on the hillside with the pups playing, and the gray wolf with the yellow eyes haunted me like a phantom. Nature can be both beautiful and brutal!

There was a new Lamar Valley pack, called the Lamar Canyon pack, and they had made a kill the day before. I drove to where the carcass was supposed to be. As expected, a large group of wildlife paparazzi was in attendance. Struggling with the flu, I staved off chills and a headache and walked to the cordoned-off area, where three rangers kept about twenty of us wolf-watchers far away from the wolves. The wolves were going back and forth to the kill site, about five altogether. Two were dark gray, and three were a mixture of white and tan. One of the dark gray wolves seemed to be dominant. He moved slowly, and it seemed that the other wolves gave deference to him and would not go near the carcass until he had left.

The wolf pack and the kill were pretty far off, so I tried to get closer by driving down the roadway to another pullout, but a ranger shooed me away, citing the potential of more cars lining the road where I had stopped. I returned to the cordoned-off corral and made the most of it. I left at sunset and settled into my hotel in Gardiner.

The next morning I headed out to the kill site. The rangers' cones were there, as were a few people, but the kill site was picked clean. A few ravens flew in and pecked at the bones.

The wolves were gone, and for the remainder of the trip I had no other wolf sightings. Still, I was grateful for the wolf encounter I did have and for my unwavering insight to "stay with the animal." I also came away with some unique wolf images.

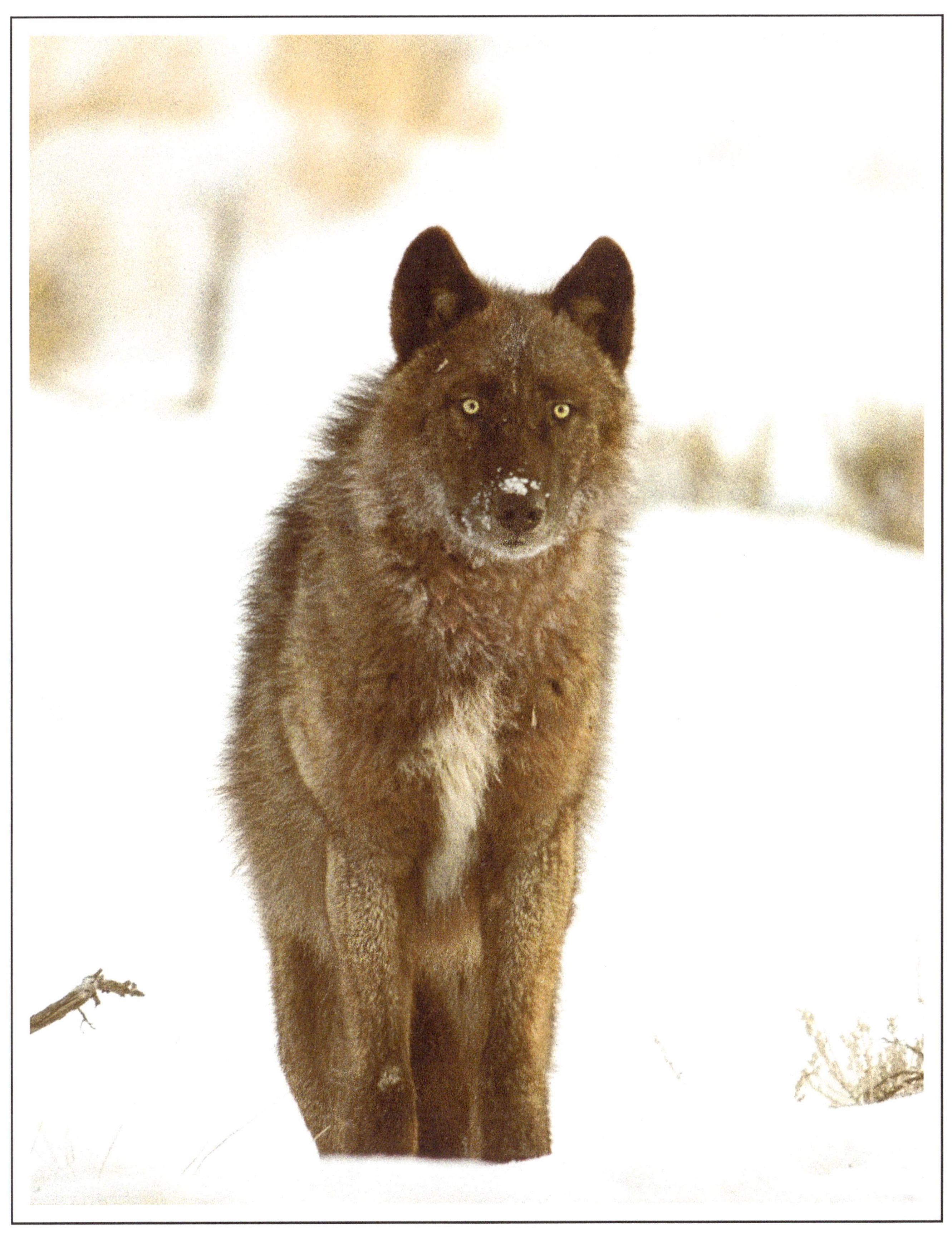

Gray Wolf "His yellow eyes locked on mine when I paused from taking photographs for an instant. Then he slipped across the road and down into the ravine." (page 67) Yellowstone National Park

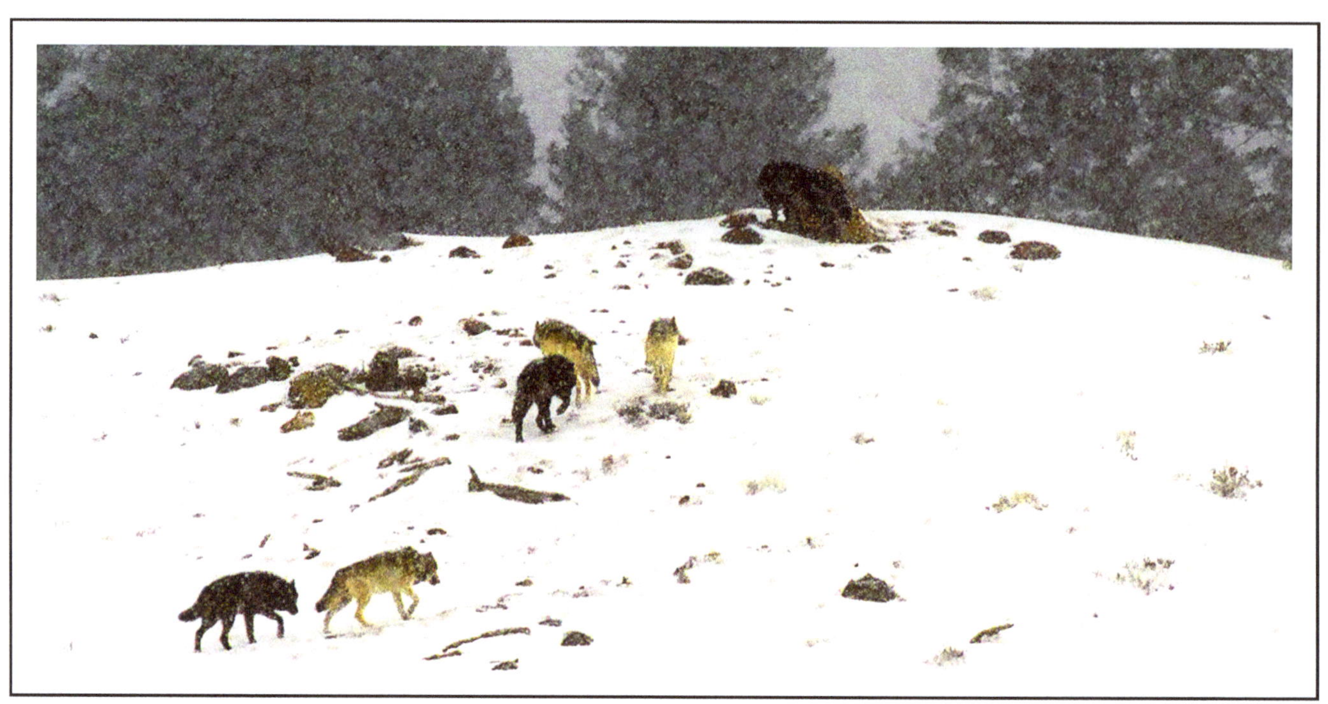

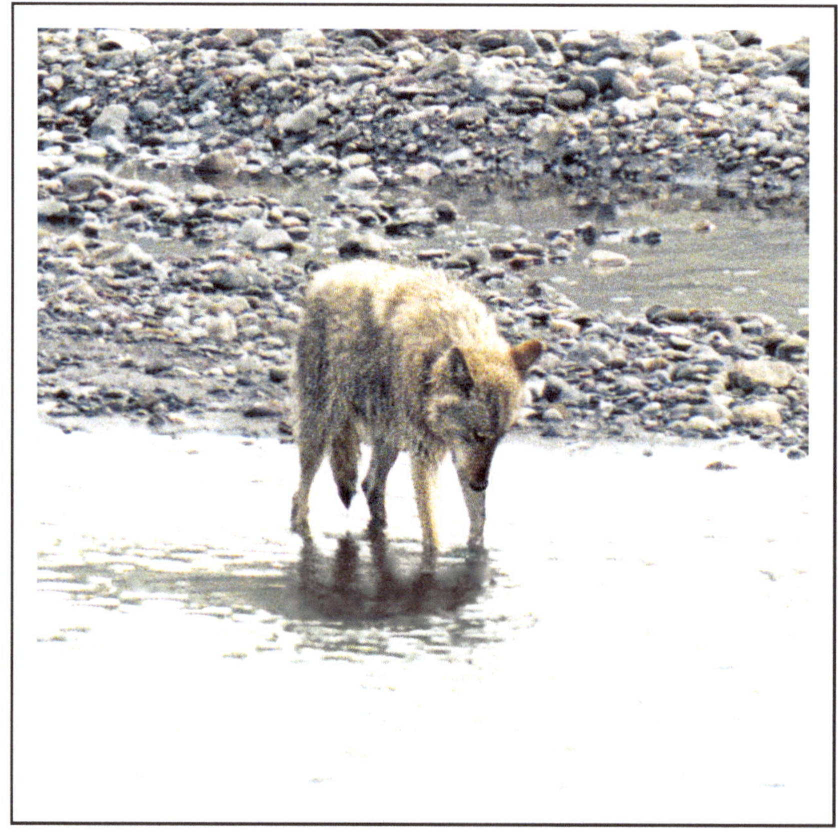

TOP: *Wolf Pack in Snow* "The weather was alternating between storm-heavy snowflakes and an occasional flurry, but I captured some shots of the wolves on the move." (page 68) Yellowstone National Park
BOTTOM: *Wolf in Stillness* Lamar Canyon Pack, Yellowstone National Park

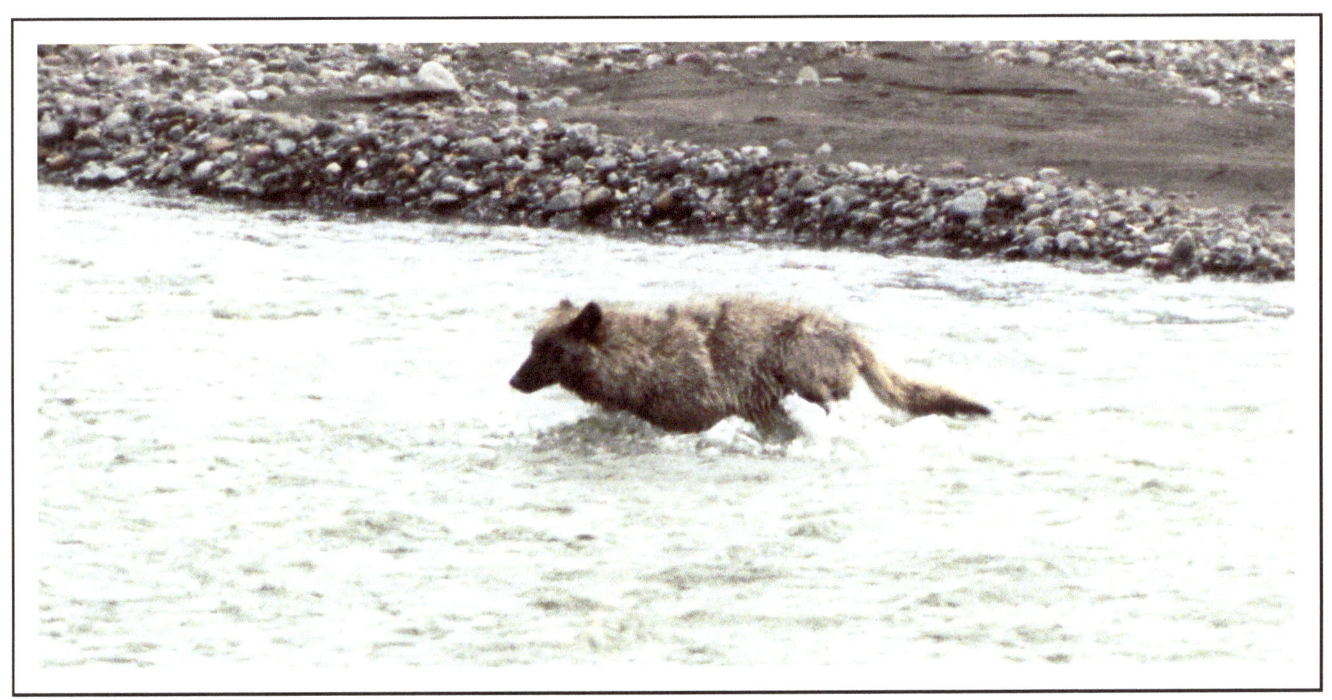

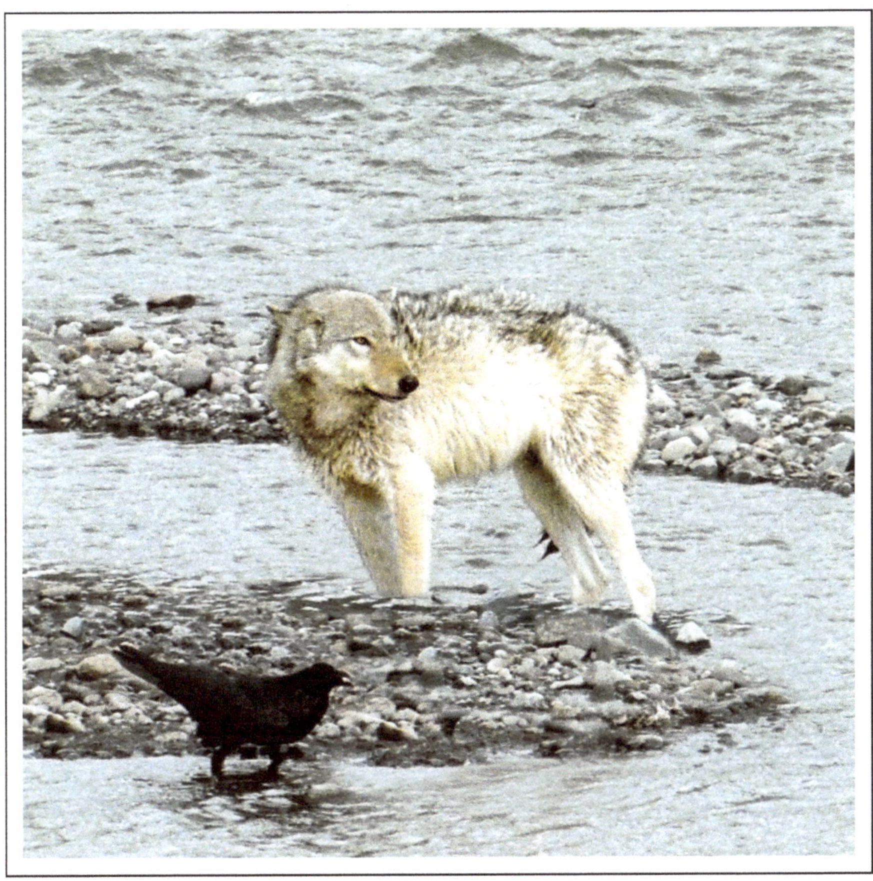

TOP: *Stream Crossing* A Lamar Canyon wolf, Yellowstone National Park
BOTTOM: *Contented Gray Wolf* "The wolves were coming back and forth to the kill site, about five all together." (page 70) Yellowstone National Park

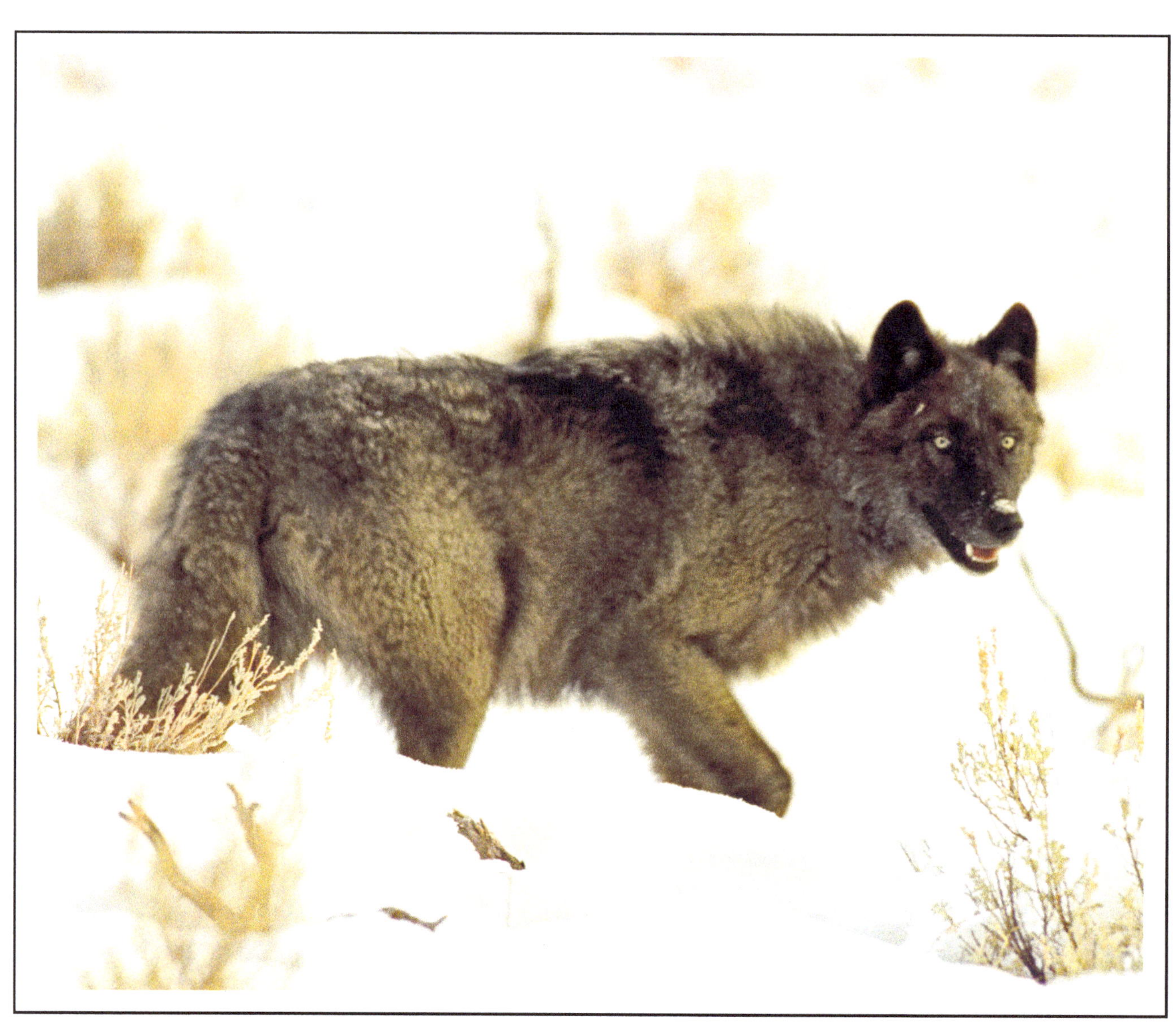

Yellowstone Gray Wolf Druid Pack, Yellowstone National Park

CHAPTER 7

Buffalo, the Usual Suspects

I was sitting on a snowmobile, facing a line of buffalo spread across a snowbound road in Hayden Valley in Yellowstone National Park, alone, confused, and starting to get nervous. The buffalo stood staring at me as though they were cemented in place. I was about one hundred feet from them. They were huge, from six hundred to a thousand pounds. They were not moving, and I didn't have room to turn my snowmobile around and go back in the direction I came.

To my right was a seven-foot wall of snow created by a snow-plow; to my left, across the narrow, one lane road was a large mountain. The buffalo and I were effectively engaged in a standoff. To make it worse, I replayed in my mind the stories I had heard from a previous Wyoming Gallery manager, Celeste Dennis, with whom I've had my work since 2000. She frequently shared with me the buffalo encounters that had occurred that year—the bison that had charged cars or the people who had gotten too close, sometimes resulting in horrendous injuries, even death.

The man I rented the snowmobile hours before had said, "If you get charged by the bison jump on the other side of the snowmobile." But I was looking at a line of seven buffalo. What side of the mountain would I jump on? It seemed a ridiculous thought—each way was equally unsafe. I began to pray for a solution.

My previous buffalo encounters had all been easy compared to this. Usually, they involved pulling over to the side of the road, standing outside the car and using a camera on a tripod to take telephoto pictures at a safe distance. Other times, I was able to photograph them from the comfort of my vehicle.

I did have a closer encounter one time while hiking in the backcountry in Yellowstone. I was on a trail with my two teenage nephews when we came across a large buffalo just eight feet from the trail not far from Yellowstone Lake. I could tell the buffalo was bothered by our presence because he had his tail up in the air. I told my nephews we would make a big circle around the buffalo and walk close to the lake. I cautioned them to not stare at the buffalo, and if they saw him starting to charge, they were to jump in the lake. Fortunately, we didn't have to swim for it—we skirted past him without incident.

This time was different: I was now trapped in a snowmobile-buffalo standoff. This was the first time I had traveled to Yellowstone in winter. I had looked forward to renting a snowmobile, because I thought it would be a unique way to see wildlife in the park. (At this time in Yellowstone in 2001, you could drive a snowmobile unguided along the plowed roadways, a policy that has since changed.) At the rental store, I had filled out the paperwork and verified my snowmobile experience. The staff member gave me a quick orientation to the snowmobile, and I was ready to go. My trip to the Cascades in Canada looking for mountain lions from a snowmobile gave me further confidence on the Yellowstone terrain.

I was determined to go to Hayden Valley, which was forty miles from the rental store at the park's south entrance. This was where the bison herd foraged for food in the winter. But what I didn't realize was how much longer it took to cover the distances in a snowmobile and how cold it would be. I don't know why, but I thought it wouldn't be as cold as it was in the Cascades. I was wrong. It was even colder. The other thing I realized is that your thumb gets really tired pressing the gas control on the handlebar of the snowmobile. After half an hour my thumb was throbbing, so I hooked a Velcro strap around the gas lever at the speed I wanted. With my snowmobile 'cruise control', I flew past many other snowmobilers on and off the road. I felt like a snowmobile NASCAR race driver, although I was very careful to not exceed the suggested speed limits, not wanting to run into any wildlife. I traveled so far I had to refuel in the canyon gas station rest area.

After refueling in the canyon area, I proceeded back to Hayden Valley. That's when I came upon the line of buffalo across the road. Like me, they seemed unsure of what to do next. They looked back and forth at one another. Again, I entertained the possibility of turning around, but the very heavy snowmobile had a wide turning radius, and I wasn't ready to leave the machine there and hike for miles over the frozen landscape.

I patiently waited five minutes, ten minutes, fifteen minutes; the situation remained the same. They were not moving, so I did the only thing I could do: I pulled out my camera and started photographing. Through the lens the buffalos' deep furrows on their faces were lit and accentuated by the ice and crusted snow that hung off their fur. The steam from their breathing seemed to create more ice as I watched. I thought, *They are such hardy animals.* There is a quiet dignity to them. They are noble creatures with a sorrowful past. Once the dominant mammal in North America, buffalo in the early 1800's had numbered in the tens of millions. But, after the new Americans from Europe settled the country, they were reduced to one thousand or less by 1890.

As I looked out at the line-up of buffalo, the thought occurred to me, *I should try moving my snowmobile to the mountainside of the road. Maybe they don't want to go between me—a potential danger to them—and the mountain; the seven-foot, plowed embankment may seem to them like an easier escape route than the steep mountainside.* It turned out to be the right solution. I edged my snowmobile to the mountain side of the road, and the buffalo began to walk past me. They stared at me as they lumbered by, some within only an arms' length, and we got by each other without incident.

This experience made me realize how gentle and non-aggressive buffalo really are if not provoked. Their massive size can be intimidating, but the instances of buffalo charging a person or vehicle actually are few and far between.

I had to get the snowmobile back, so I fastened my Velcro cruise control and shot down the road. Yellowstone no longer allows unguided snowmobile trips. This is probably a good idea.

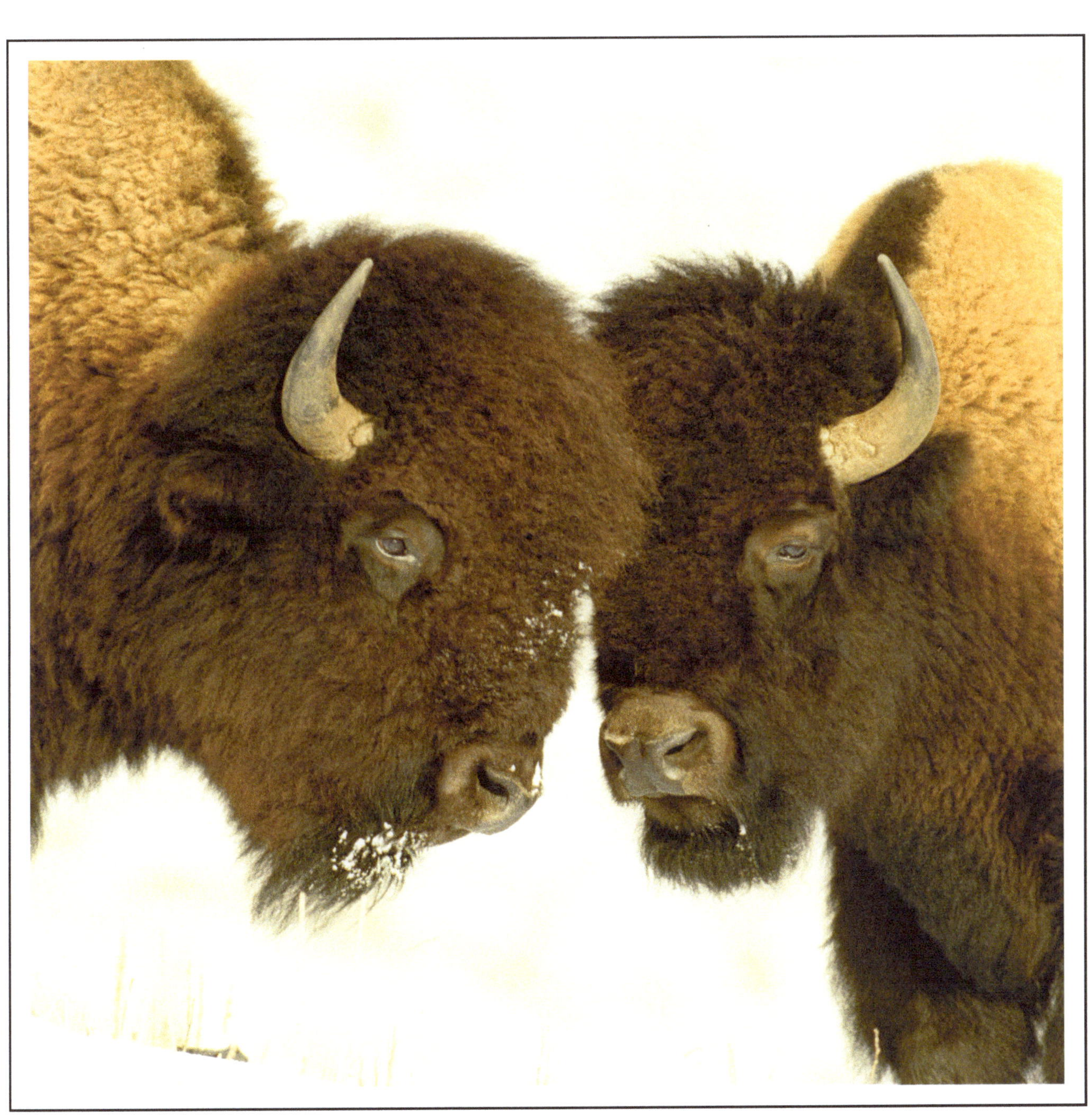

Buffalo Buddies Yellowstone National Park in winter

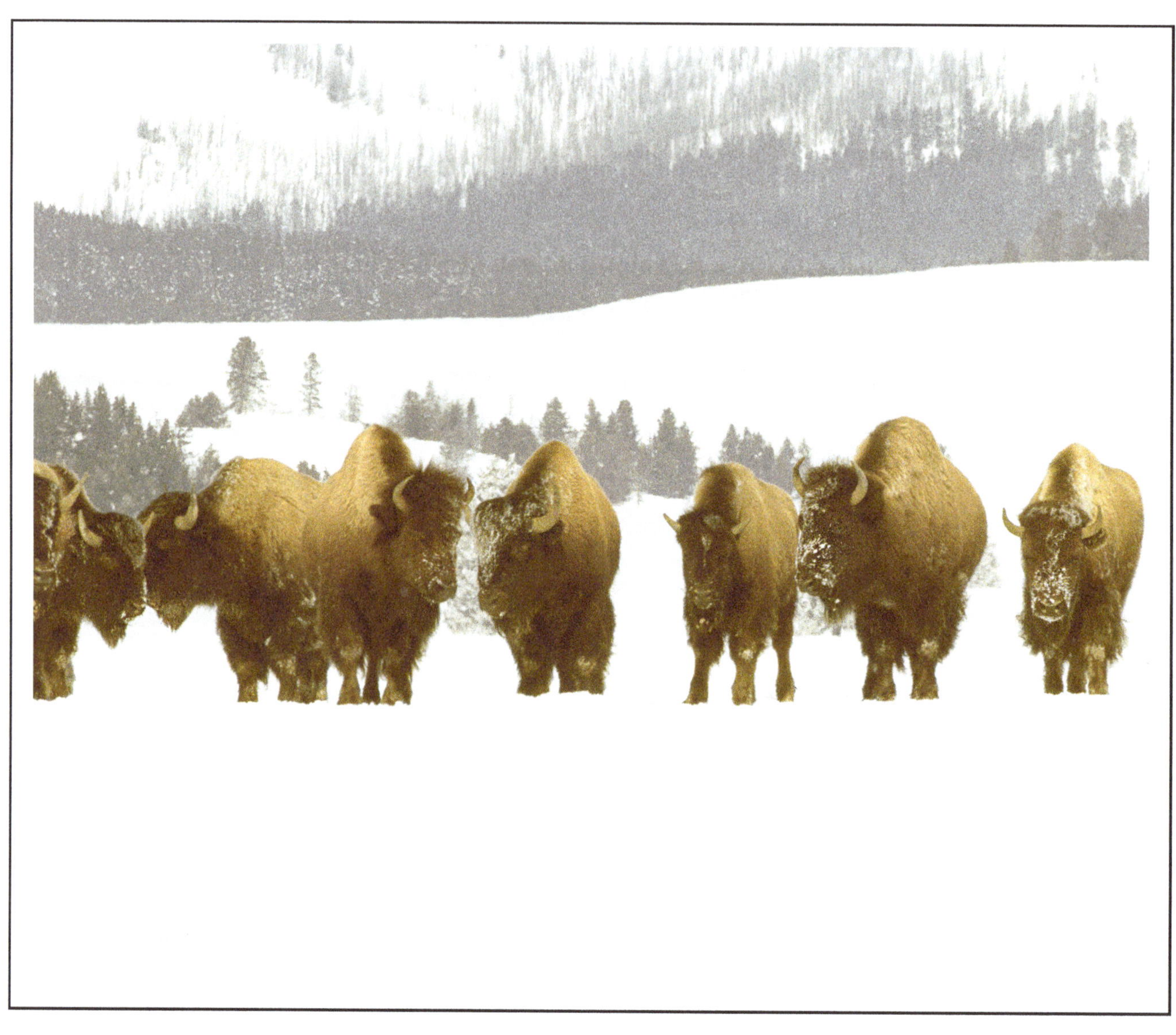

The Usual Suspects "The buffalo stood staring at me as though they were cemented in place. I was about one hundred feet from them. They were huge, from six hundred to a thousand pounds. They were not moving, and I didn't have room to turn my snowmobile around and go back in the direction I came." (page 75) Yellowstone National Park

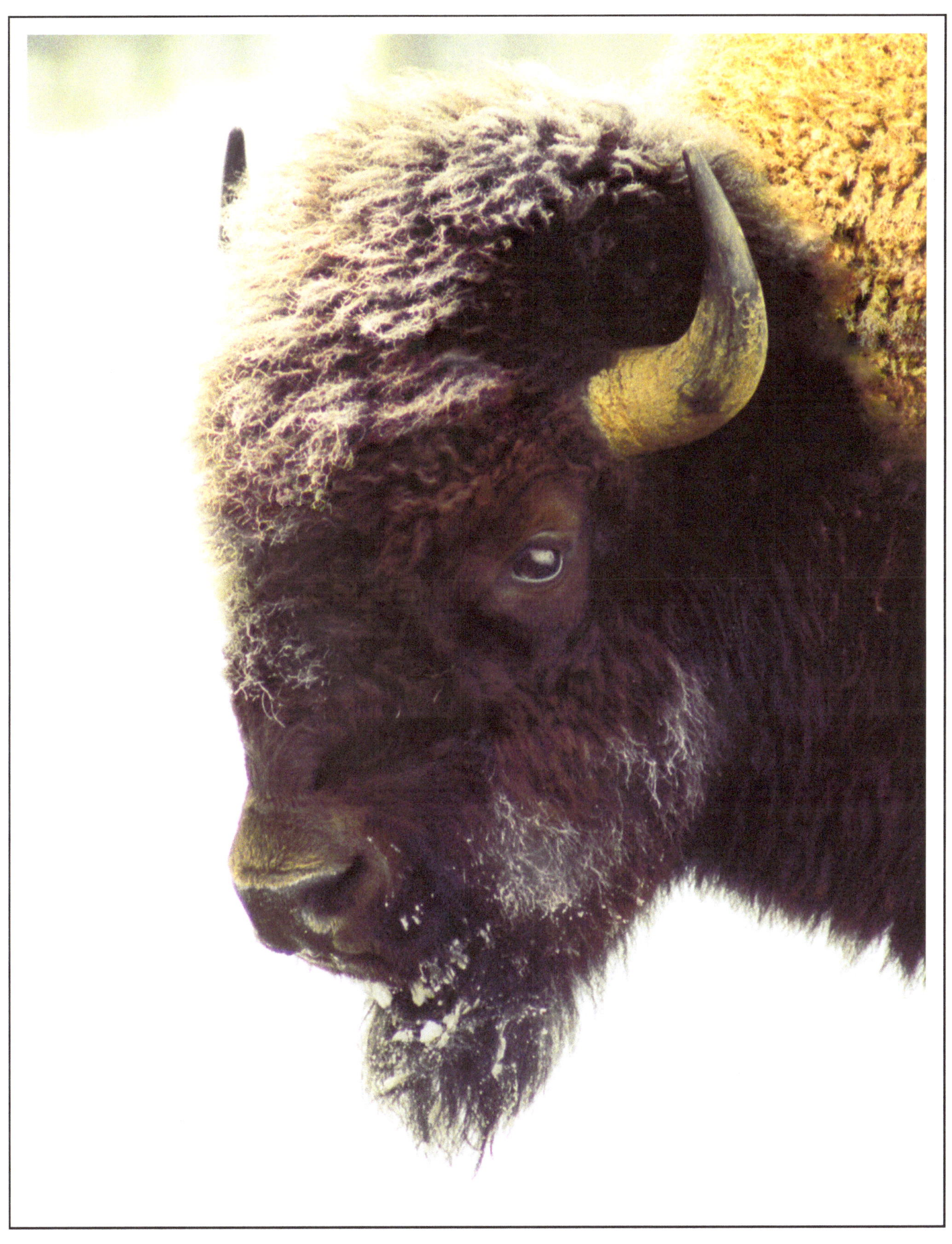

Buffalo Portrait Taken in Yellowstone National Park

CHAPTER 8

Polar Bear Adventures

I arrived in Churchill, Canada, on a small plane the second week of November 2009 after spending a night in Manitoba, Canada. Churchill is an isolated town on the shore of the Hudson Bay, a small town a little larger than twenty square miles with a population of around one thousand people. It is surrounded by Arctic tundra to the northwest and the Hudson Bay to the north. The south of the town is a boreal forest, but permafrost keeps the tree cover sparse and stunted. When you look around, all you see is flat white snow, snow drifts, and puffy snow-covered bushes. From October through December, the temperature hovers between twenty-nine degrees above zero Fahrenheit to nine below. The town is famous for the many polar bears that gather along the shore of the Hudson Bay, where they are waiting for the bay to freeze to hunt seals out on the ice. I was there in an organized commercial tour, the best way to view the polar bears.

When our group of fifteen arrived in Churchill, we were given a bear orientation that consisted of the following: if you're out walking in town and you see a polar bear, run into a building, any building. No clapping, no yelling "Hey, polar bear," no backing away slowly—just running. Not much to remember, just run! On the ice with my large, clunky boots, five pounds of clothing, and twenty pounds of photo gear this would be tough.

Our first day on the tundra started with an evening dinner trip out to see the bears. Even though it was dark, it was an amazing experience to see my first polar bear. The bears were massive, seven to eight feet tall, with heads ranging from sixteen to twenty-five inches wide and paws twelve to fifteen inches wide. The huge bears juxtaposed with the smaller people in the tundra buggies brought home the fact that polar bears view us as prey. *It would be hard to walk safely among them,* I thought.

We traveled to the bears in tundra buggies, large truck-like vehicles built with fire-engine chassis and extra-large, snow-gripping tires. The buggies elevated viewers above the bears to keep them safe. The bears seemed to be captivated by the vehicles. They would stand on their hind legs and paws and lean into the buggy. Some bears' heads would be within three to five feet of the people in the buggy.

The next day out on the tundra, we saw many sleeping bears, some younger bears playing, and mothers and cubs on the move. When large male polar bears wandered onto the scene, the smaller bears became agitated, jerked their heads, stared at the larger bear, and then ran quickly in the opposite direction. It was clear the polar giants were feared and respected.

The bears seemed captivated by the tundra buggies and the smells coming from the lunch being served on them. Chicken cacciatore was a popular lunchtime favorite.

I kept thinking the bears must think humans taste like Italian food. They also seemed to like the salt on the bottom of the buggies. Whatever drew them closer, it all made for great photos and videos.

The third day out on the tundra was filled with many cubs and their polar bear moms. I was amazed at what good mothers they seemed to be. The cubs crawled all over their moms who stoically tolerated it. The polar bear moms seemed frequently on the move, as larger male polar bears sometimes kill the cubs to bring the polar mom back into estrus, or mating readiness. Sometimes this threat appeared to contribute to the moms' stress.

The only time I felt a little anxious was when I was in the town of Churchill. The night before our last day, a young polar bear came into town, killed a sled dog tied up at a house, and then had to be put down. The normal bear-deterrent procedure in Churchill was to have the towns' bear patrol shoot off fire crackers and fireworks to scare away stray bears. The bears that ignored the noise were darted and held in "Polar Bear Jail" until the ice froze. They were then helicoptered out to the ice. Lately, many polar bears had been wandering into town and the Bear Jail was full. While I was there, the ice wasn't frozen; I was told it froze later, in December. But the Bear Jail was so full they were helicoptering bears back to the tundra to make room for the more recent offenders.

On my last morning in Churchill, I woke up early to take a walk and photograph the town. I came upon a large Native Canadian statue and a tour bus of people taking photos of the sunrise and the Hudson Bay on the outskirts of town. I was busy taking videos myself, when I noticed the tour bus had left. I was alone and about a long city block away from town. With the beautiful sunrise on the shore of the Hudson, I stayed another fifteen minutes enjoying the view.

The silence was shattered by fire cracker noises seeming to come from over the hill to my right. I realized quickly I was in danger—I had wandered far from any buildings or stores. The thought of a polar bear running at me over the hill clicked me into panic mode. I grabbed my camera gear and ran as fast as my clunky boots could slip and slide over the icy roads back to town.

Once I was safe in town, near a large medical complex, I reflected on my stupidity for wandering out of town, the guide's voice ringing in my head: "Do not go out of town." Seeing all those people by the statue had drawn me there, and I realized how easy it is to be sloppy and forgetful. I wonder how many Churchill residents forget and do what I did, but with bad consequences. A polar bear walking through town seems like such a foreign concept to me. But in Churchill this occurs annually during October through December, when the polar bears gather to go out on the ice.

Later that day I was able to go out to the Bear Jail where they kept the bears that had wandered into town. The jail looked like a large airplane hangar. As we came around the side of it, two rangers had a heavily sedated polar bear with her cub in netting waiting to be lifted off. A helicopter flew in, attached a hook to the netting, then lifted the bears up in the air and started flying off towards the distant tundra. It felt like it was the end of a *Born Free* movie, and all of us watching cheered the bears' new fate. *What a strange trip it must have been for the polar bears when they awoke on the tundra*, I thought.

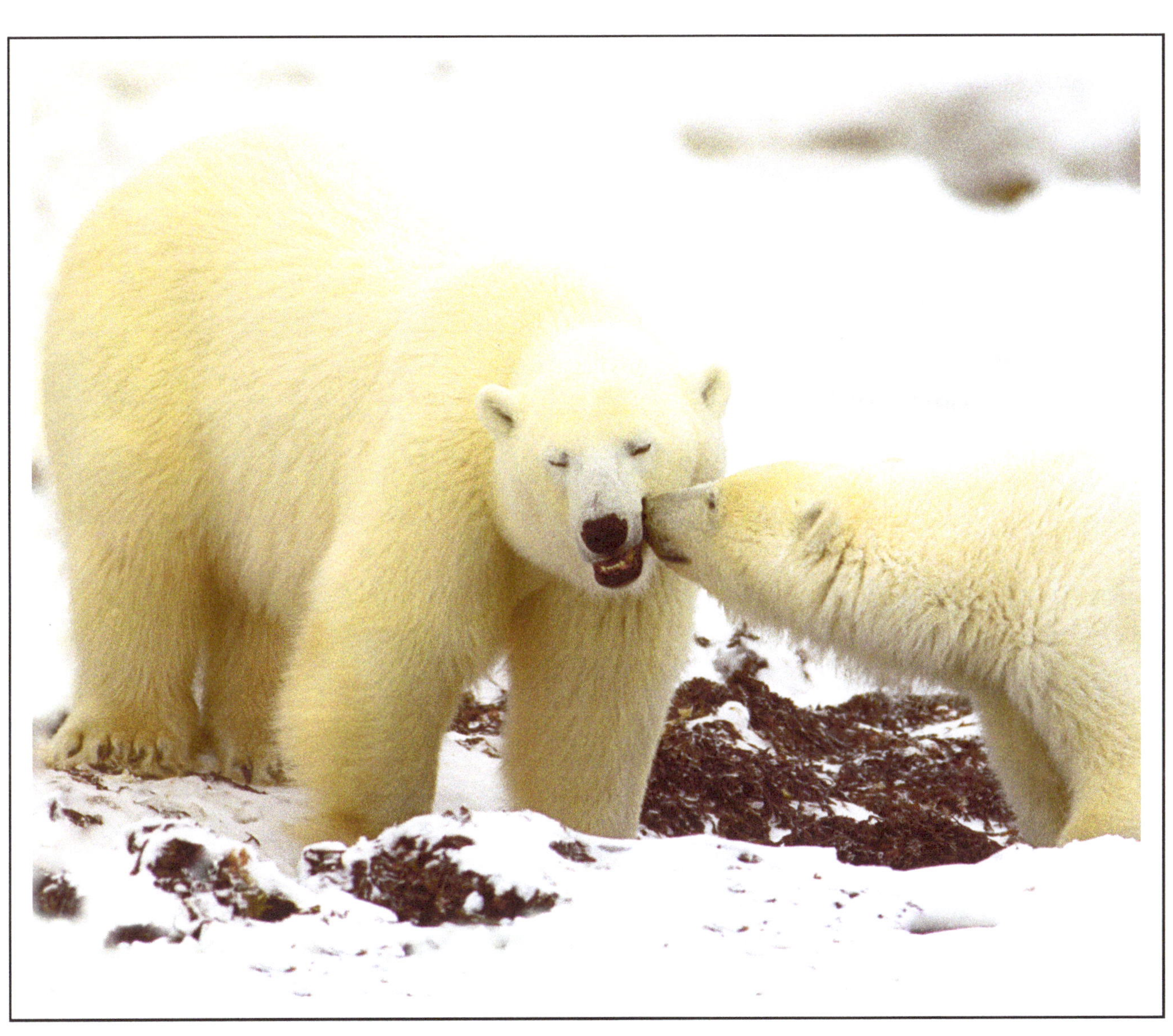

Polar Mom Churchill, Canada

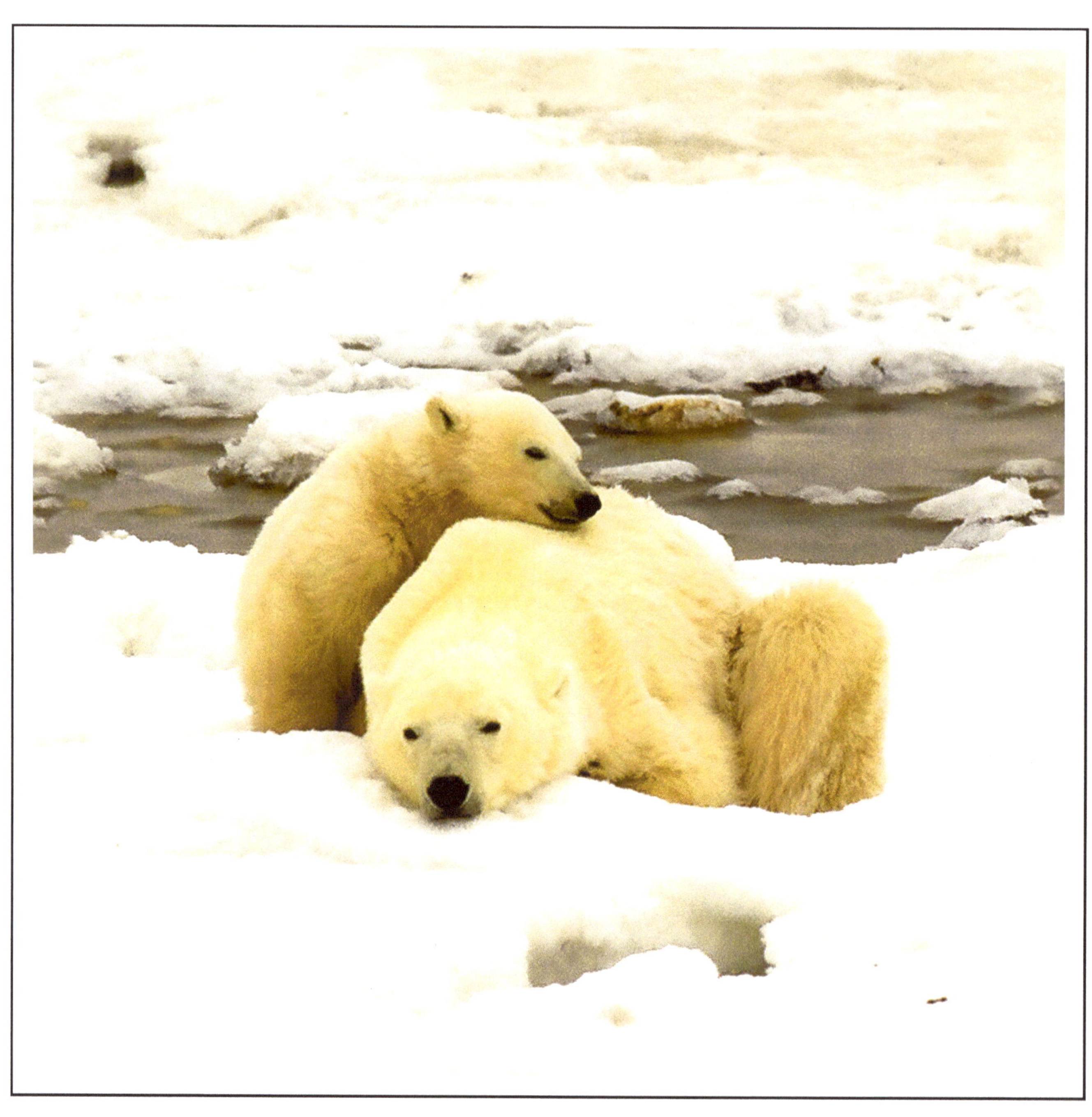

Polar Nap "I was amazed at what good mothers they seemed to be." (page 81) Churchill, Canada

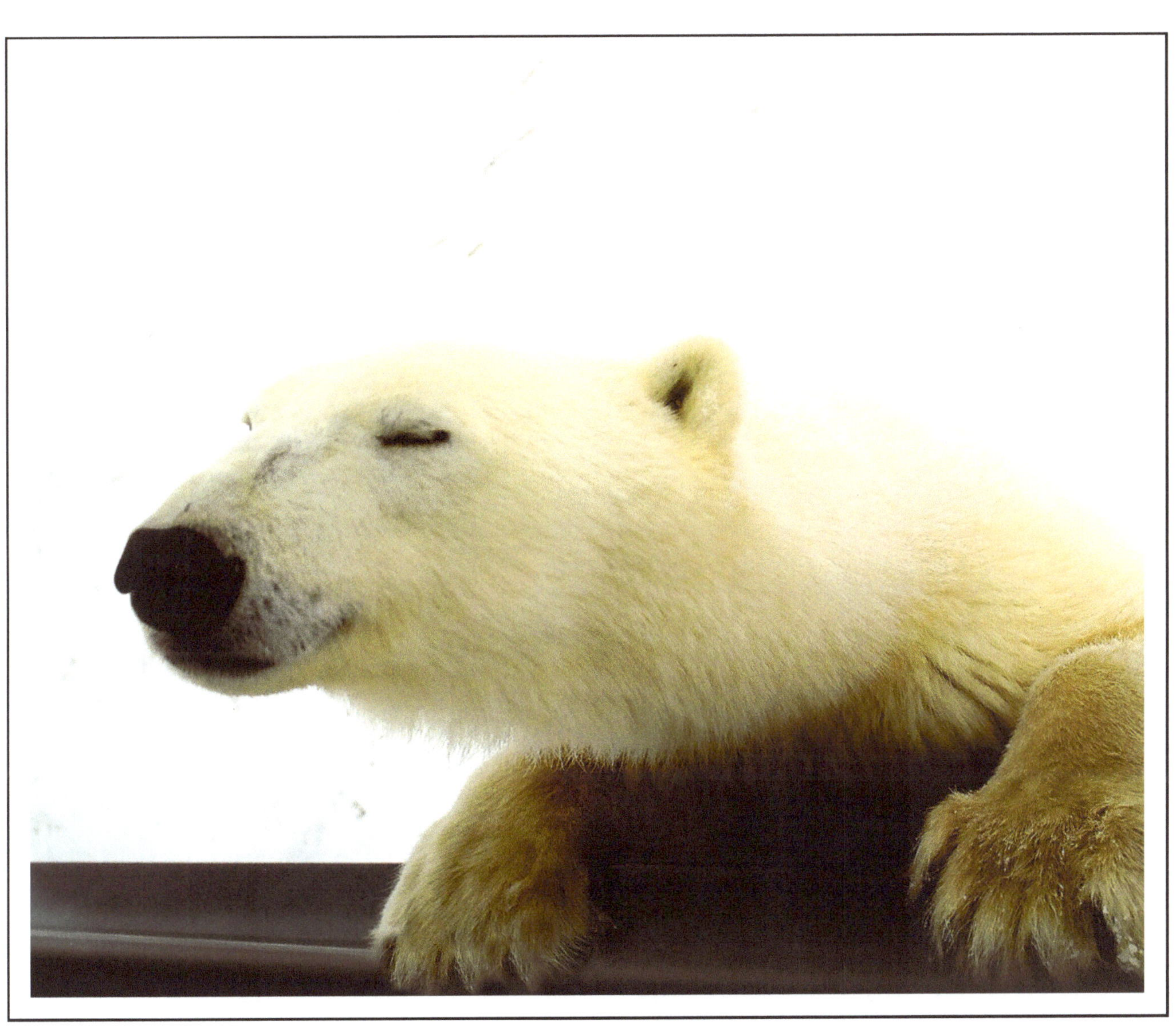

Polar Breeze "They would stand on their hind legs and paws and lean onto the tundra buggy. Some bears' heads would be within three to five feet from the people in the buggy." (page 80) Churchill, Canada

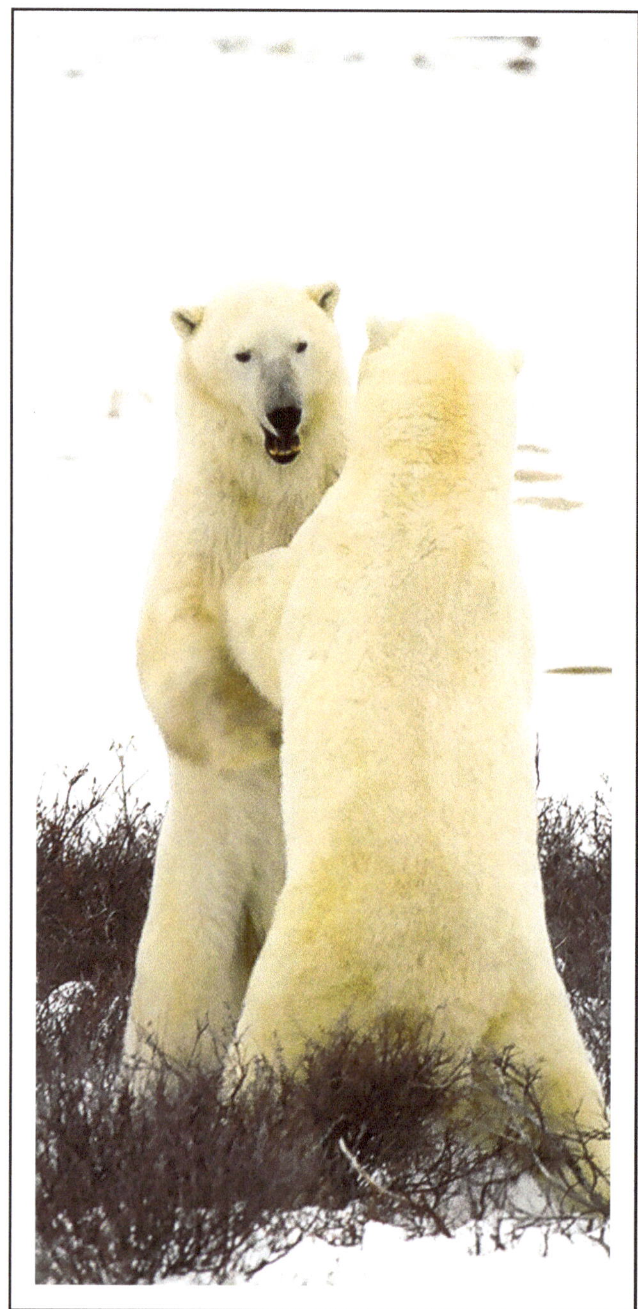

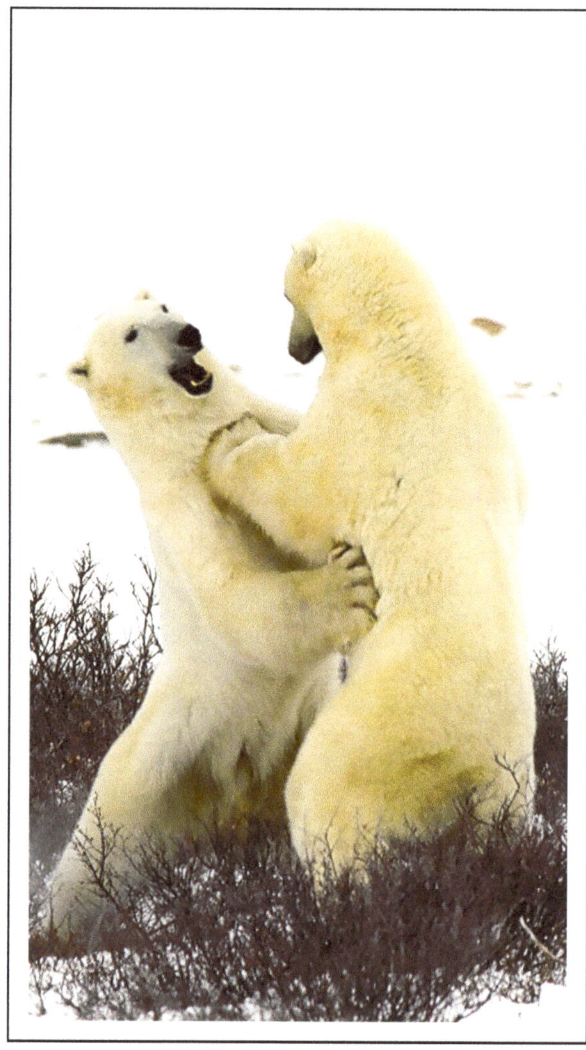

LEFT: *Polar Greetings* (image one) These two young polar bears were jostling with each other while waiting for the ice to freeze.
RIGHT: *Polar Greetings* (image two) Churchill, Canada

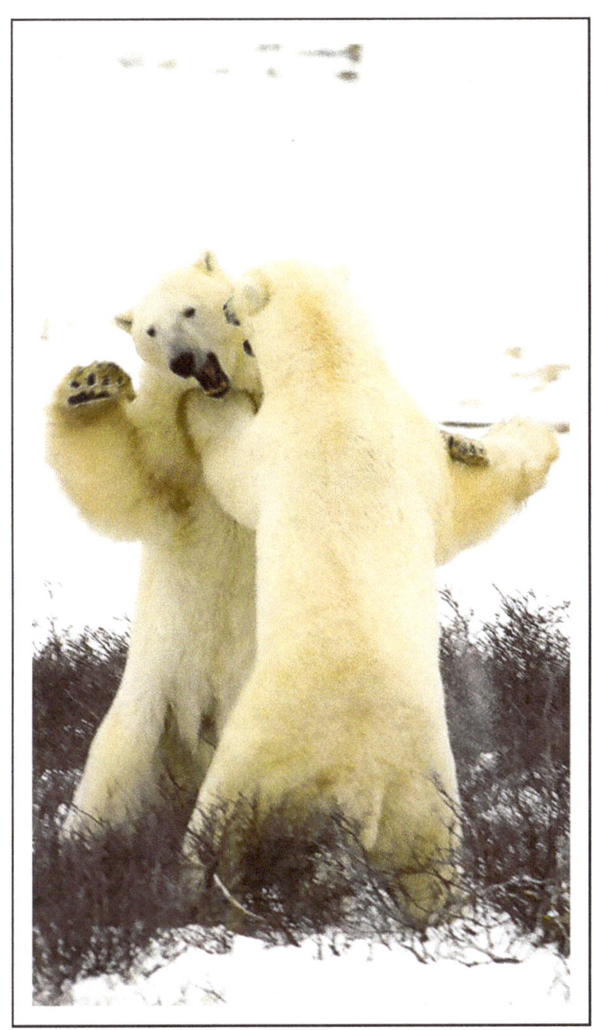

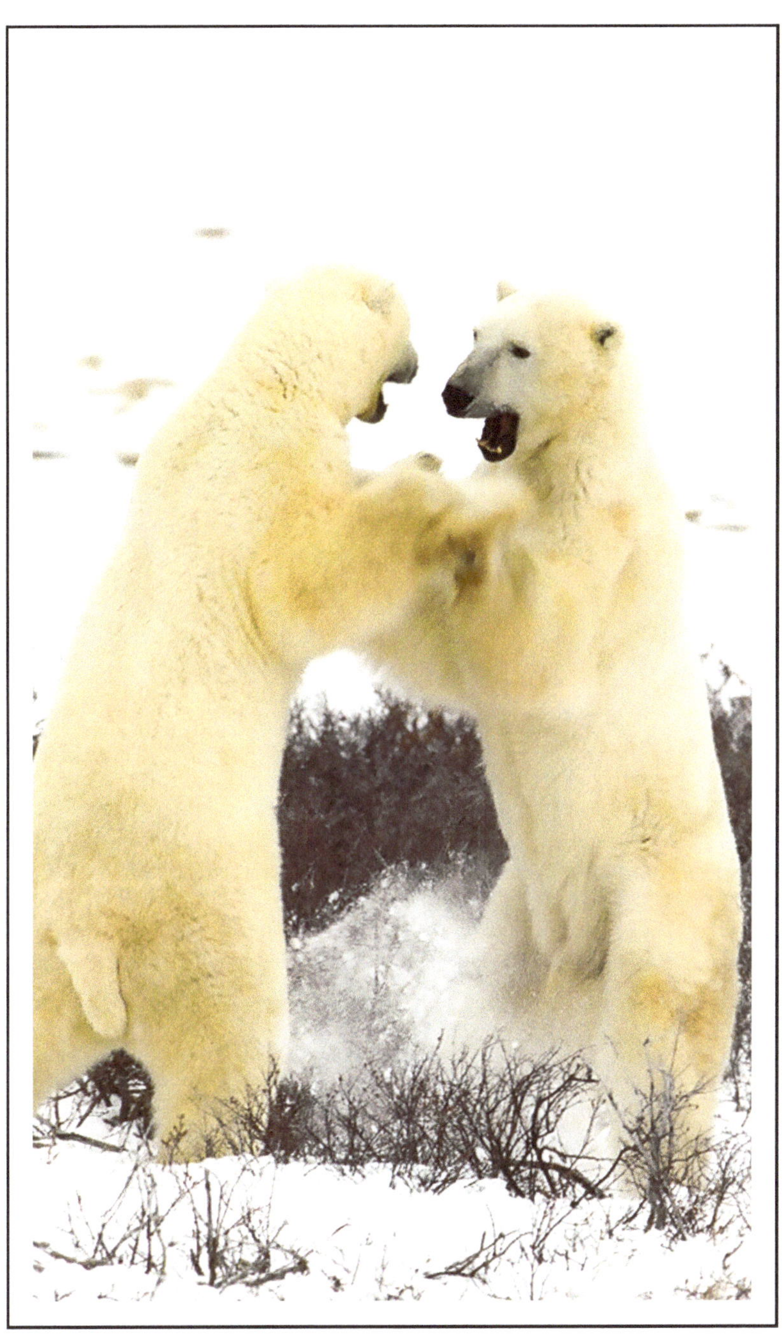

LEFT: *Polar Greetings* (image three) Churchill, Canada
RIGHT: *Polar Greetings* (image four) Churchill, Canada

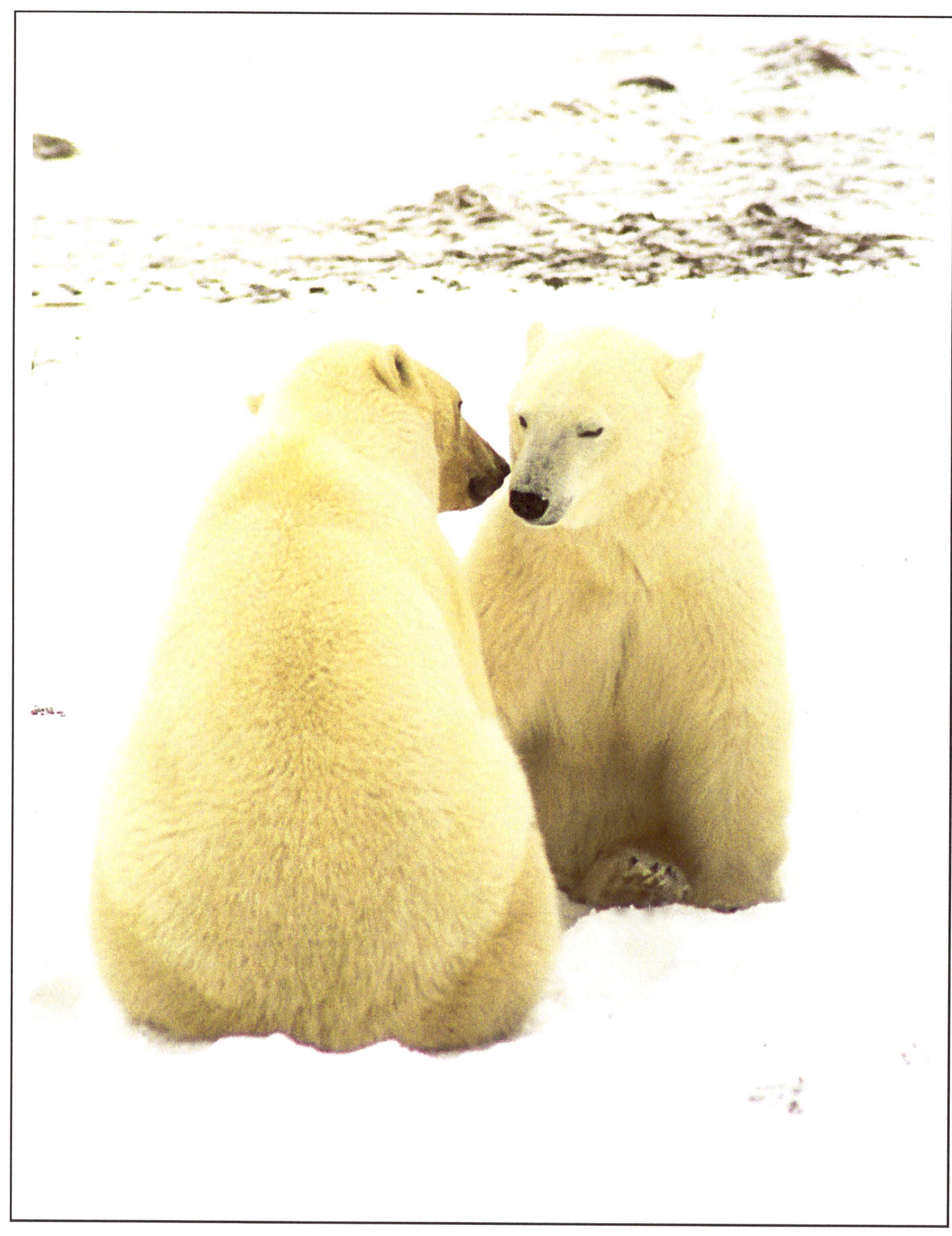

Polar Affection Churchill, Canada

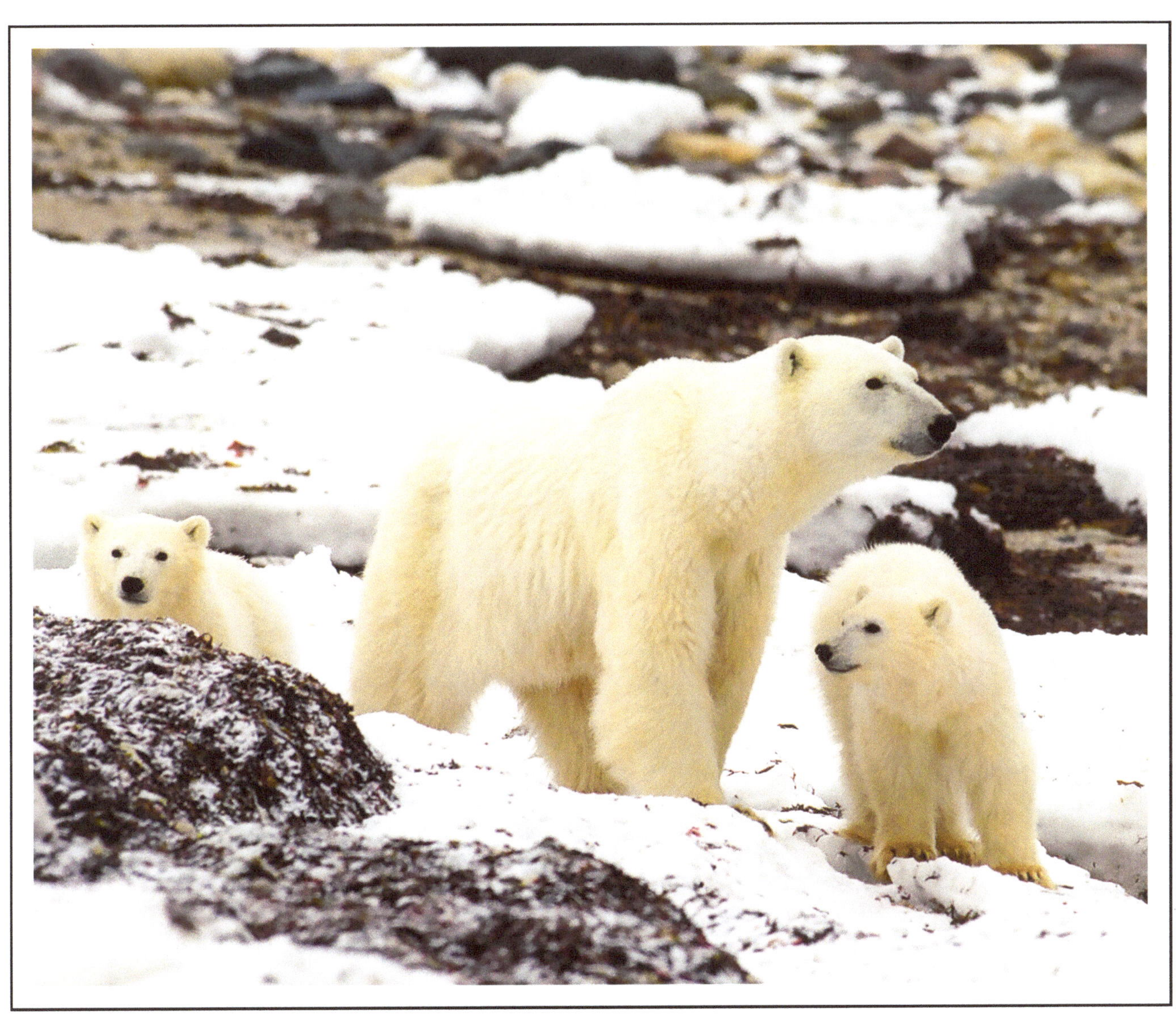

Polar Bear Family "The polar bear moms seemed frequently on the move as larger male polar bears sometimes kill the cubs to bring the polar mom back into estrus, or mating readiness." (page 79) Churchill, Canada

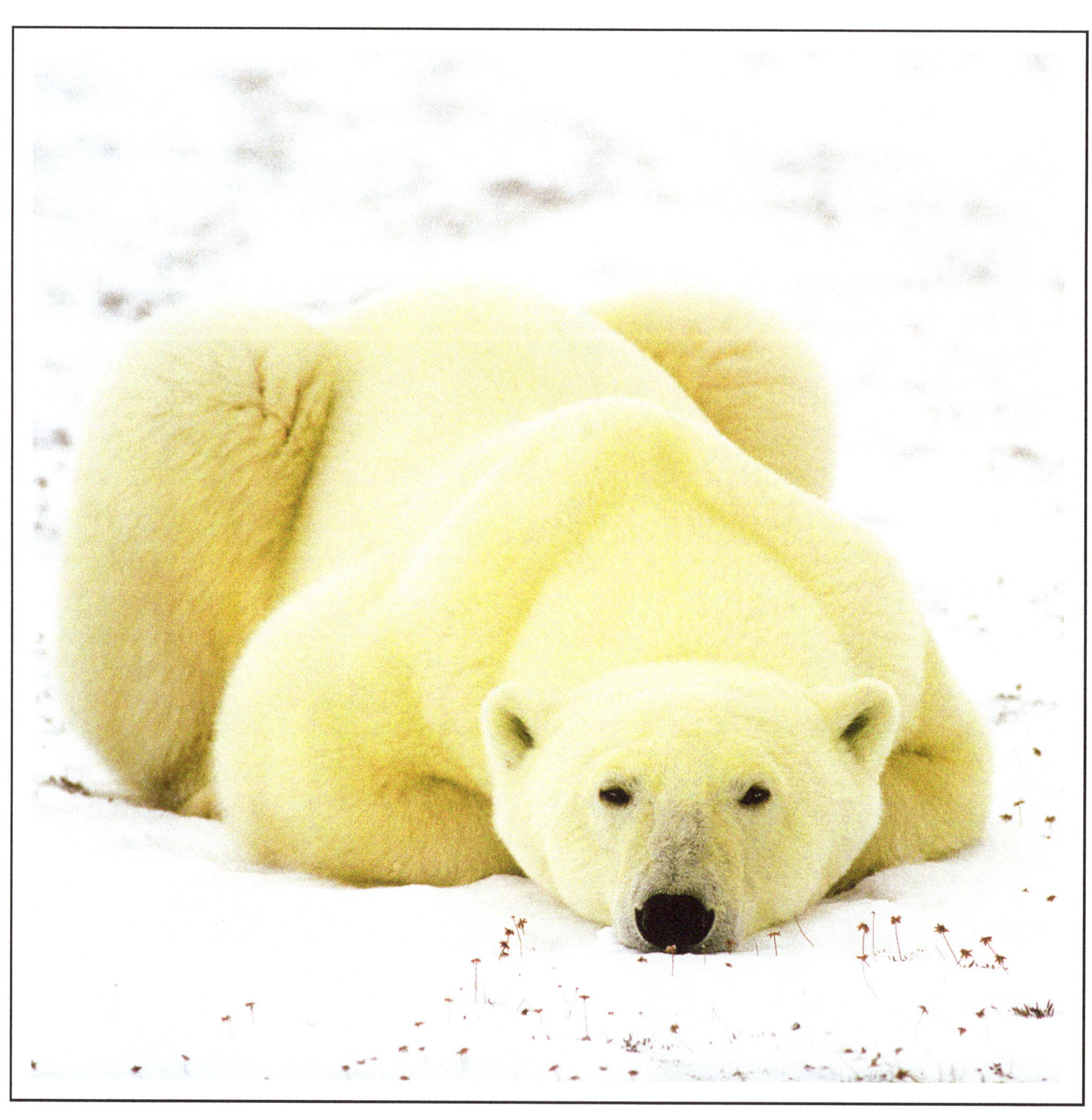

Polar Morning "The next day out on the tundra we saw many sleeping bears, some younger bears playing, and mothers and cubs on the move." (page 80) Churchill, Canada

CHAPTER 9

Wild Horses in America

It was early morning. The sun was creeping through a thick mist rising off a small creek lined with trees. I stood alone, about forty feet from a small group of wild mustangs, a "harem of horses," so-called because it was led by a dominant stallion. In addition to the main stallion, there were one small foal, three mares, and two young stallions. The relationship between the small foal and the three mares protecting him was tender. He would rest his head on the back of one of the mares while she ate grass. The mares encircled him and I imagined this was how it was when many herds of wild horses dotted the landscape, a by-gone time in the American West.

Captivated by the scene, I was surprised when the dominant stallion came at me, reared his head, snorted, and showed his teeth as he whinnied. Trembling and holding onto my camera gear, I got the message clearly and backed away. Given a wide berth, he and the others settled back into their routine. This behavior was similar to other herd behavior I had experienced with elk in Yellowstone. I have always backed away slowly when confronted by an aggressive male or female animal. There is give-and-take: if I give them plenty of room, they seem to tolerate my presence.

The lighting was beautiful that day, and I was making the most of it with my camera. As I got some great shots, I reflected on my first experience with wild horses.

I had always been fascinated by the reports of wild horses still running free in America and wanted to see and photograph them. With their strong muscular legs, flowing manes, and remarkable speed, horses are like gentle giants. They are large animals who can easily cause serious injury with their kicks. But after they are broken, they frequently develop close and protective relationships with humans.

My first encounter with wild horses occurred quite accidentally. While on a trip to Deadwood, South Dakota, with my husband and young daughter, as I was reading about activities in a guide book, I read about a herd of wild horses in southern South Dakota. Excitedly, I signed us up for a guided tour to the Black Hills Wild Horse Sanctuary, near Hot Springs, South Dakota. Founded in 1988 by rancher Dayton O. Hyde, the sanctuary is on eleven thousand acres of private land. The wild horses were obtained from the Bureau of Land Management (BLM) which probably would have destroyed them if they weren't adopted out. The horses at the sanctuary run free with very little intervention.

We rode out to see the wild horse herds in an old school bus with a seasoned tour guide. It was midday in the summertime, so the lighting wasn't the best for photography, but I hadn't expected to get any good photographs zipping along in the bus anyway. The bus driver took us into the center of a large harem of horses that numbered at least thirty. Surprisingly, two large stallions were engaged in a brutal fight. As the larger of the two jumped away, I was able to get an amazing shot of him, despite the moving bus. The other horse had a deep, large tear in his skin on the side of his abdomen from a bite, and he quickly ran off. The guide said the sanctuary didn't intervene in the daily life of the wild horses, and that fights, though uncommon, could be deadly. It was strange to see horses engaged in this type of behavior—it seemed so harsh and un-horse-like, but it also was exciting, wild and made me more curious about the way the horse herds functioned.

I subsequently went looking for wild mustang herds in Nevada. More than half of all wild mustangs are found in Nevada on public lands, and their presence has not been without controversy. Their supporters feel mustangs have a right to the land, as any wild animal does, that they are part of our national heritage. Those who oppose this policy feel that horses degrade rangeland and compete with livestock and other wild species for vegetation. In 1971 the U.S. Congress passed laws to protect the horses and recognize mustangs as living symbols of the historic and pioneer spirit of the West, which continues to contribute to the diversity of life-forms in the nation and enriches the lives of the American people. The wild mustang herds are managed by the BLM.

With some research, I learned that many of these wild horse herds were found around Pyramid Lake, a large saltwater lake in the Nevada desert, and I planned a trip to photograph them. On the way, I stopped at a BLM horse adoption center where they had just finished a roundup of a large number of wild horses from public lands. They were trying to adopt them out. The BLM representative was gracious and told me I could go out to see the horses in their corrals.

The dominant stallions were separated from the foals, mares and younger stallions. Fifteen to twenty large males were together in the stallion corral. I heard in the distance the whinnying of a horse, which I assumed was being broken so the horse could be ridden. One of the large stallions in the corral whinnied back. The distant horse whinnied again and the stallion in turn communicated back. And so it went, with the stallion getting agitated, shaking his head, and kicking at other horses. It was very disturbing to see. Maybe these two horses were family members being forcibly separated. I left the area with tears in my eyes, feeling their bond on a primitive level. I wish there was an easy answer for the wild horse situation in America.

Continuing on my way to Pyramid Lake, I stopped for gas and lunch at a roadside café. It was packed, and an enthusiastic young teen at the counter asked if I was heading to "Burning Man." After looking around, I realized many vehicles were going in one direction on this small desert road. Unknowingly, I had scheduled my wild-horse-stalking trip to coincide with the "Burning Man Festival" which was taking place in the same area where I had hoped to find wild mustang herds.

In this long snaking, conga line of vehicles were VW pop-up vans, uniquely designed homemade RVs, and new and old cars all loaded with people, camping gear, and coolers.

For those who don't know what 'Burning Man' is, it is a large festival where for one week each year over 50,000 people descend on Nevada's Black Rock Desert. According to the Burning Man website, it is a festival dedicated to community, art, self-expression, and self-reliance. The event ends after a huge wooden replica of a man is set on fire during a ritual—hence the name Burning Man. From what I hear, it's unique, kind of like a twenty-first-century, art-focused, Woodstock Festival without the big rock bands.

I presumed that with all these people, the noise, and construction there wouldn't be a wild horse within a hundred miles, so I headed south.

Not far from Pyramid Lake, I went to a Native American museum and spoke to an employee about specific locations where I might find wild mustang herds. He said they usually followed creeks and small rivers and that I might try searching north of the Reno area.

The next day, while staying in Reno, I woke up before sunrise and set out on interstate 80 going east. Outside of Sparks, Nevada, I saw a small river running parallel to the freeway and remembered what the man said about creeks and rivers. I took a side access road that seemed to skirt along the river's edge, but after what seemed like miles it stopped abruptly, and I was forced to return to the freeway. Another five miles down the road, I picked up the river again, and an access road appeared. I followed the river with its bends and turns.

I was almost ready to give up when, out of the corner of my eye I saw a small herd of wild mustangs by the creek. I parked in a small pullout and hiked up to a grassy embankment about forty feet from the horses. This was the mustang group whose dominant stallion had driven me back from his harem, as I described at the start of this chapter. It was one of my most amazing wildlife encounters, and I stayed with the small herd for about half an hour until they moved on.

My last encounter with wild horses occurred on the East Coast at Assateague Island, an Atlantic coast barrier island thirty-seven miles long. The northern two-thirds of the island are in Maryland,and the southern third is in Virginia. Assateague Island National Seashore is best known for its herds of feral, free-roaming wild horses of domestic ancestry. They are viewed best in the Maryland portion of the park.

Assateague's wild horses were different from the herds I saw in Nevada and South Dakota. They seemed smaller, and many had white or blond manes bleached by the sun. They reminded me of beach bums. The sands of the island were white, with scattered grassy dunes, on which the horses fed. Since they were accustomed to people, the horses hardly took notice of me—they sauntered around undisturbed. Compared to the herds in Nevada and South Dakota, Assateague's horses were more scattered about.

I wasn't adequately prepared for this trip. I had a new digital camera that wasn't functioning well, and the persistence of swarms of biting flies made it hard to be outside my rental car. Because my camera wouldn't work, I missed a rare opportunity to photograph the amazing sun-bleached blond horses running along a white sandy beach, their manes flowing in rhythm, as they galloped past me. I resolved never again to go on a photo trip without my old, manual-focus, film camera. With all the advantages of digital cameras, I still like to have my old metal-housed film camera. It could be dropped, the meter could malfunction, and yet it would still work.

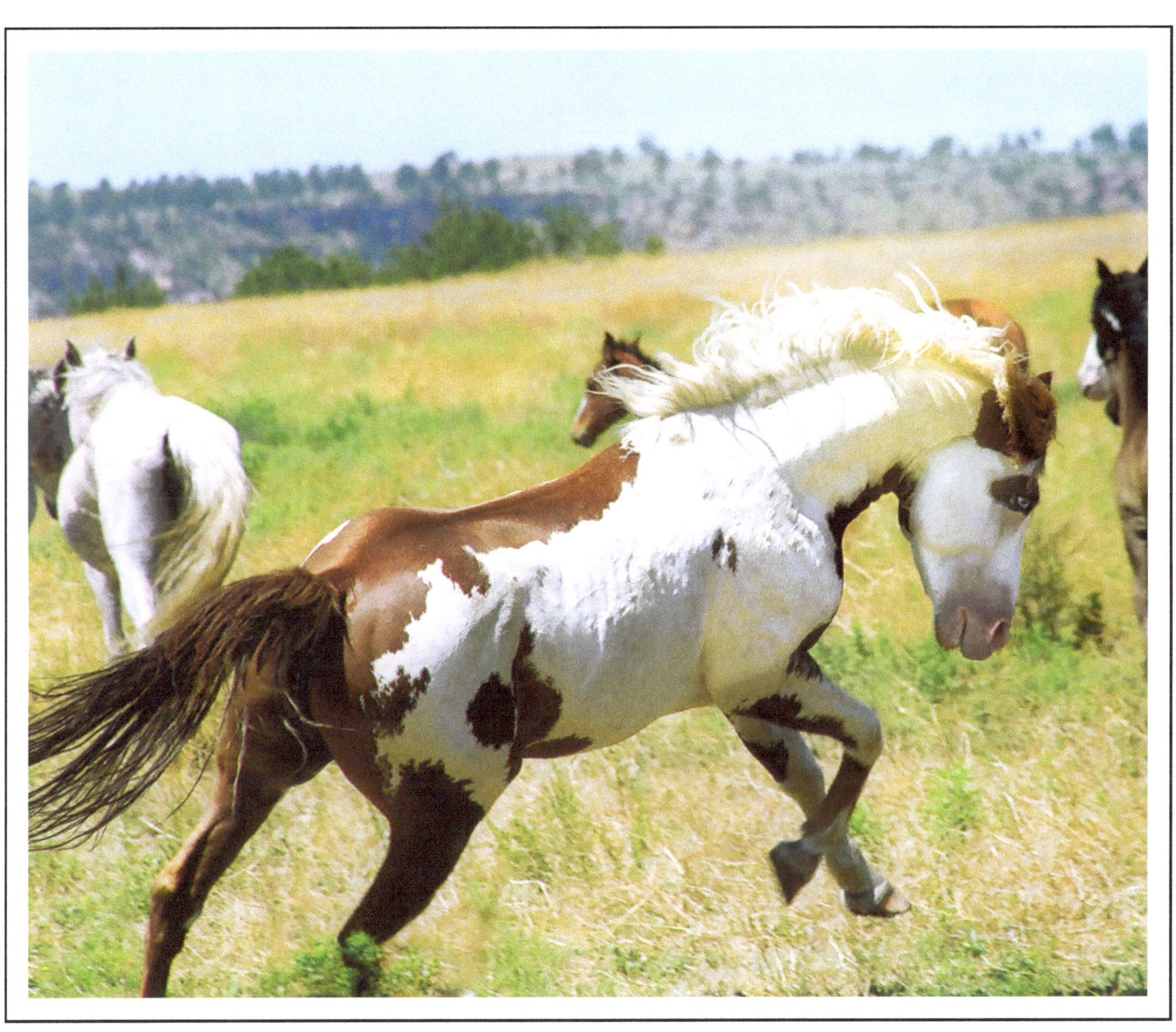

Wild Horses "Surprisingly, two large stallions were engaged in a brutal fight." (page 92)
Black Hills Wild Horse Sanctuary, South Dakota

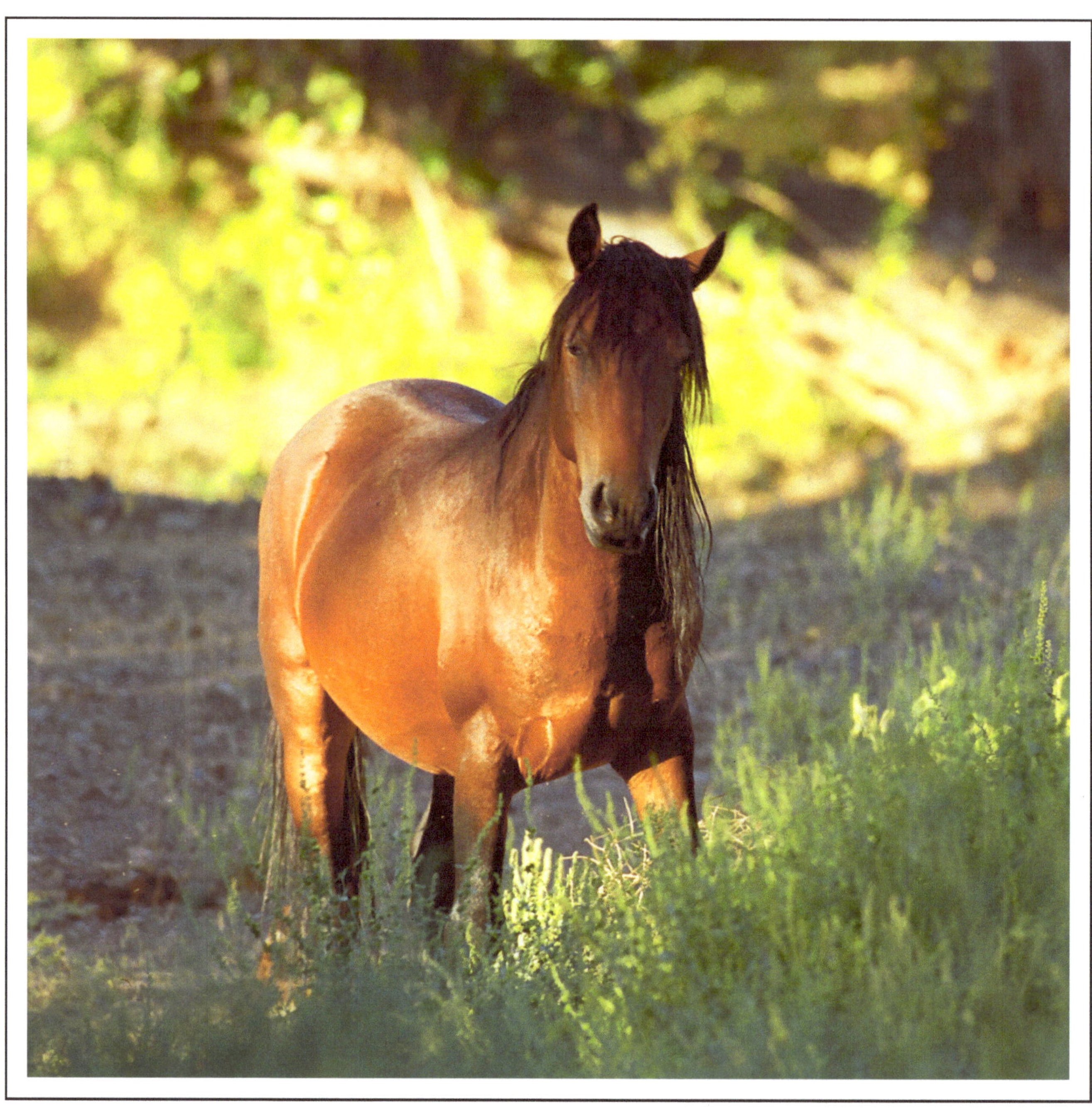

Wild Mustang Stallion "I was surprised when the dominant stallion came at me, reared his head, snorted, and showed his teeth as he whinnied." (page 91) Nevada

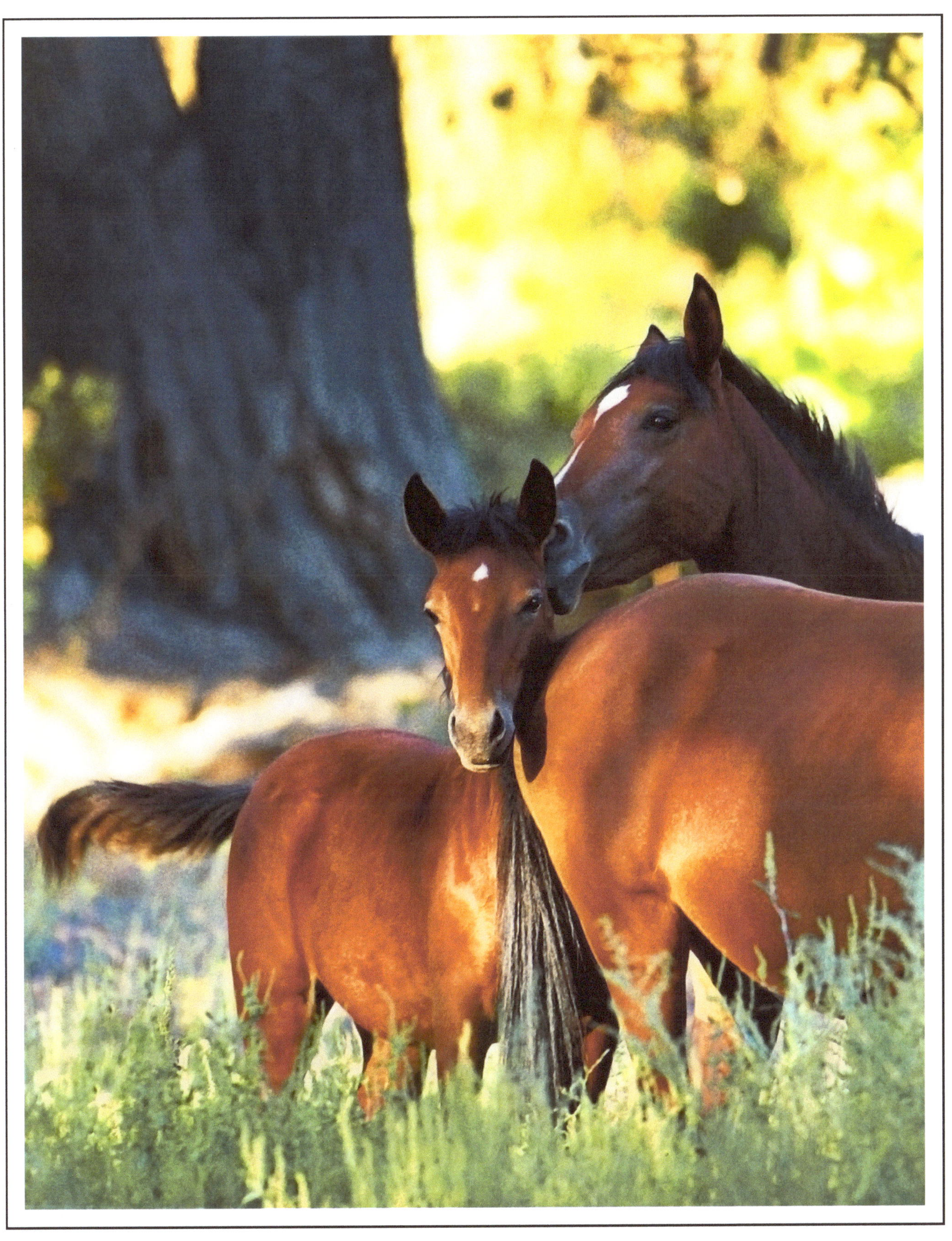

Wild Mustang Family "The relationship between the small foal and the three mares protecting him was tender. He would rest his head on the back of one of the mares while she ate grass." (page 91) Nevada

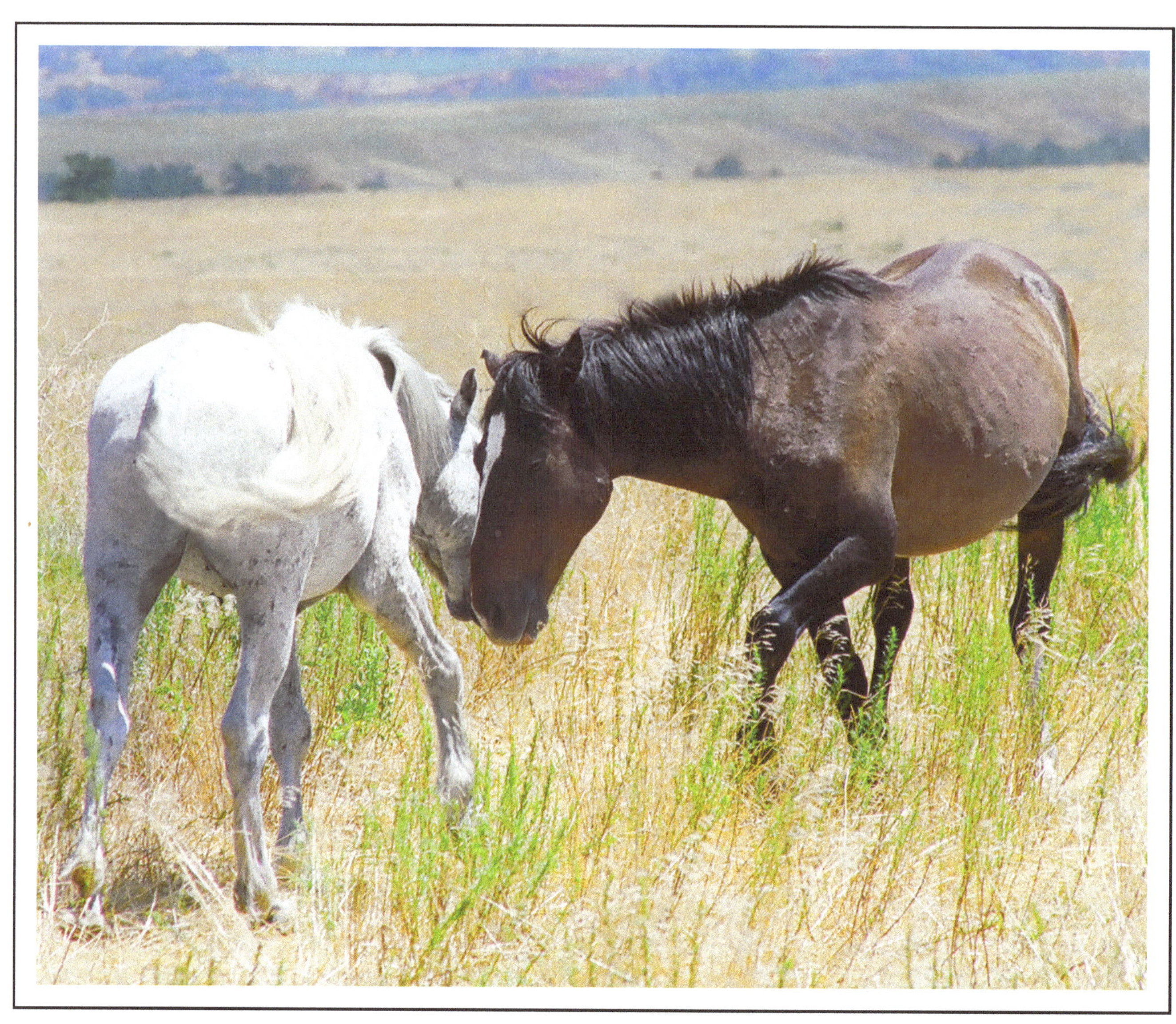

Wild Horse Romance Black Hills Wild Horse Sanctuary, South Dakota

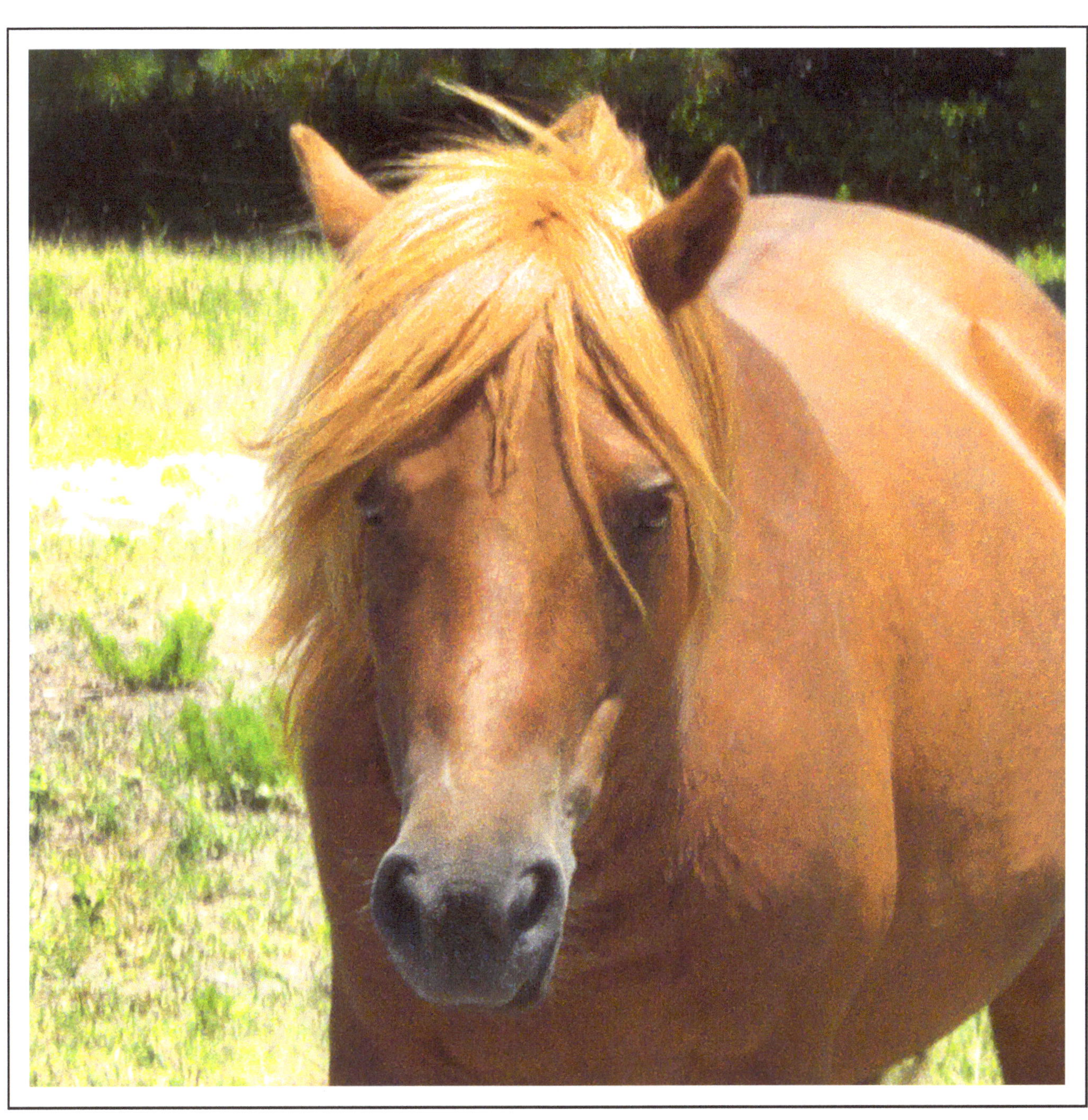

Sun-Kissed Wild Stallion Assateague Island, Maryland

AFTERWORD

All of the wildlife images in *Wild Among Us* were taken in the wild in North America. The only alterations were the removal of radio collars from the mountain lions and from two wolf photographs for aesthetic reasons. The *Black Bear Gathering Acorns* is a composite of four images scanned together to show all of the stages of the black bear's activity. I also use the terms grizzly bear and brown bear interchangeably in Chapter 2.

While on photography trips, I never forget that wildlife are wild and can be dangerous. I try to give them the respect they deserve by not intruding on their space. I would rather abandon a shoot if the animal is too stressed, than continue agitating the animal. In this regard, I always try to do what the authorities recommend in places where wildlife roams. By reading the wildlife warnings that are posted and/or dispensed by the rangers I acquire valuable information on how to be safe within that park or wilderness area.

When I reflect on my time photographing North American wildlife I can't help but think of these elusive animals as symbols of our connection to the world around us. The degree to which these creatures are healthy, beautiful and thriving is a measure of how our planet is functioning as a whole. When we hear about black bears, or mountain lions wandering into towns and cities because of drought conditions in the high country, polar bears going out on the ice later and later and having a shorter seal hunting season, and grizzly bears and other wildlife struggling because of reduced salmon runs, it all reminds us of our fragile ecosystem—that our actions do matter—and that we impact each other. The more we can live lightly on our planet, the longer we can continue to live among, love, and admire these great animals for many years and generations to come. In North America, we are all privileged to have these wild animals among us.

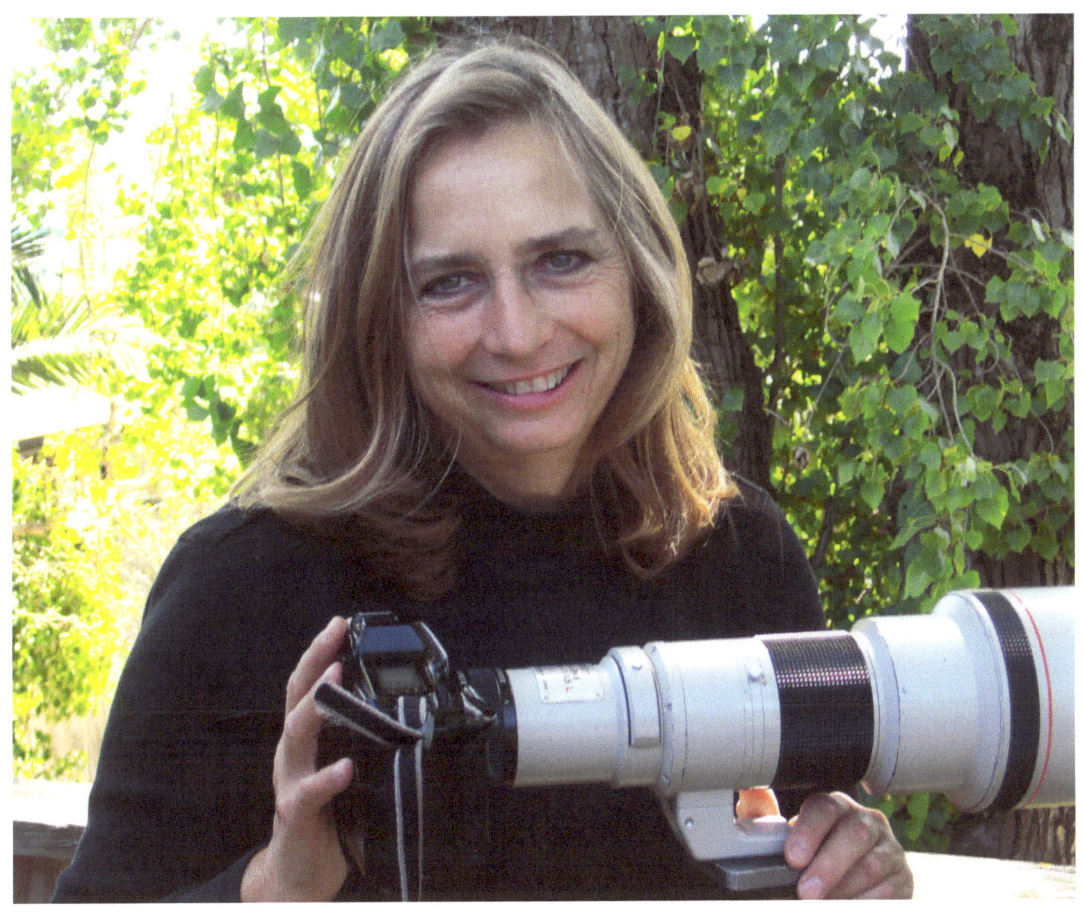

ABOUT THE AUTHOR

Pat Toth-Smith is an accomplished wildlife photographer with over twenty years of professional experience. Her bear and other wildlife photographs have sold in numerous galleries throughout the West, and in hundreds of fine art shows. A featured artist in *Wildlife Art* magazine in 2005, Pat also was a guest artist at Yosemite's Ahwahnee Hotel in 2008, and has sold her work in Yosemite, Sequoia and Kings Canyon National Parks.

The concept for the book, *Wild Among Us*, arose out of the frequent questions her customers asked: Where was that taken? How close were you to that bear? How did you, a woman, get into a career like this? She noticed when recounting the sometimes harrowing, always exciting, encounters she had with wildlife to get a certain image, that her encounters seemed to personalize the images, making them more meaningful and memorable.

www.ingramcontent.com/pod-product-compliance
Lightning Source LLC
LaVergne TN
LVHW071631100826
845154LV00007BA/132
9780989251334